Knowledge Quest

The World of
Science
and
Technology

Is this curious-looking machine
a stunt prop from a movie or
an actual solo flying machine?
Find out on page 146.

Knowledge Quest

The World of
Science
and
Technology

Published by
The Reader's Digest Association Limited
London • New York • Sydney • Montreal

Contents

How to use Knowledge Quest

Knowledge Quest: The World of Science and Technology is a uniquely interactive reference book that brings you the essential facts and a wealth of wide-ranging information on science and its application.

Knowledge Quest will make adding to your store of knowledge both interesting and fun. It builds into a highly illustrated reference series, with each volume delivering authoritative facts and many other significant things to know about a particular branch of knowledge.

The World of Science and Technology is packed with facts on the various branches of science and their application in the world around you – all contained within the core reference section (pages 39 to 153). If you want to use the book as a straightforward source of information the contents page lists the major topics covered, while the detailed index (starting on page 161) allows you to look up specific subjects. If you simply like to browse, you will find that each fascinating piece of information leads you to discover another, and another, and another . . .

What's so special about Knowledge Quest?

The unique feature of **Knowledge Quest** is the set of quiz questions that can be used as an entertaining way to get into the reference information. You can use these to test out what you already know about science and technology, and to lead you eagerly into finding out more.

How do the questions link to the reference section?

There are 100 quizzes of 10 questions each, which are graded and colour-coded for levels of difficulty (see right). Each question is accompanied by a page number, which is the page in the reference section where you will find both the answer and more information on that subject generally.

The answers to all questions relating to the topic of the spread – which will come from several quizzes – are listed by question number in the far-left column. A number (or sometimes a star) following each answer refers you to the box containing the most relevant additional information elsewhere on the spread. More than one box may be indicated, and sometimes none are especially relevant, in which case an additional detail is given with the answer.

The questions in each quiz lead you to several different pages and topics in the reference section, which is great for finding out more, but is not so convenient if you want to use the quizzes as straightforward quiz rounds. So, for all keen quizmasters, we also list the answers to all the questions in each quiz in the 'Quick answers' section which starts on page 154.

Each question in the quizzes at the front of the book is linked by a page number (given immediately below the question number) to a page in the reference section, where you will find the answer, plus additional information on the subject. In this example, question 85 is linked to page 114.

The answers to all the questions relevant to the overall topic of a reference spread are listed by question number in the left-hand column. Each answer is followed by a box reference showing where more information can be found. Sometimes, instead of a box reference, additional information is given with the answer.

Each box on the rest of the spread contains information about an aspect of the overall topic. In this example, the topic of pages 114–115 is 'Matter' – the atoms, particles and molecules that make up solids, liquids and gases. Box ❹ features information on molecules. Sometimes, more than one box will be relevant to an answer.

From question . . .   to answer . . .   to discovering more

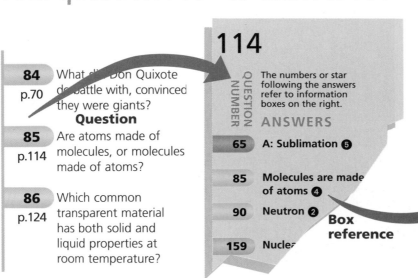

84
p.70
What did Don Quixote do battle with, convinced they were giants?
Question

85
p.114
Are atoms made of molecules, or molecules made of atoms?

86
p.124
Which common transparent material has both solid and liquid properties at room temperature?

114

QUESTION NUMBER

The numbers or star following the answers refer to information boxes on the right.

ANSWERS

65 A: Sublimation ❺

85 Molecules are made of atoms ❹

90 Neutron ❷

159 Nuclea...

Box reference

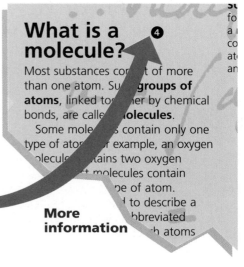

What is a molecule? ❹

Most substances con...t of more than one atom. Su... **groups of atoms**, linked to...her by chemical bonds, are calle... **...olecules**.

   Some mole...s contain only one type of ato... ...r example, an oxygen ...lecule ...ains two oxygen ...t molecules contain ...pe of atom.

...d to describe a ...bbreviated ...h atoms

More information

Colour-coding for quizzes of different levels of difficulty

A thousand questions are provided in a hundred themed quizzes. There are three levels of difficulty – **Warm Up**, **In Your Stride** and **Challenge**:

Warm Up
quizzes feature easy questions to get you into the swing. Children may enjoy these quizzes as much as adults.

In Your Stride
questions are pitched at a more difficult level than 'Warm Up' quizzes, requiring a little more knowledge of the subject.

Challenge
quizzes are the most difficult, often requiring in-depth or specialist knowledge. But have a go anyway – you never know what information may be tucked away in the recesses of your brain!

Two other categories, **All Comers** and **Multiple Choice**, include mixed-level questions, ranging from 'warm-up' level to 'challenge', and generally becoming harder as a round progresses.

All Comers
quizzes cover a range of levels of questions, from easy to hard. Everyone should be able to join in and see how they get on.

Multiple Choice
quizzes offer four possible answers to each question, only one of which is correct. They feature mixed-ability questions, generally arranged to become harder as a quiz progresses.

Special features

Star answers

One answer on each two-page spread is marked with a **star** in the answer column. This indicates a subject of special or unusual interest within the spread topic. (Occasionally, more than one answer in the answers column refers the reader to the star answer box.)

 **Biomass**

Biomass
Biomass consists of any **renewable organic material** – living or recently living plants and

Tie-breaker questions

Some spreads, but not all, feature tie-breaker questions:

Tie-breaker

These questions can be used by quizmasters in a quiz in the event of a tie, but they also contain information that expands on the answer for the interested reader.

Keeping score

The most straightforward **scoring method** is simply to award one point – or if you prefer, two points – for every correct answer.

If you are using the quizzes as rounds in a competition, you may find it easiest to look up the answers in the **Quick Answers** section at the back of the book.

For readers who would like to use this book as an information source for setting quizzes, a blank **Question sheet** and **Answer sheet** are provided at the back of the book. Quizmasters can photocopy the question sheet for their own use, and the answer sheet for distribution among contestants.

Quiz Questions

In at the beginning
All the following describe things that start with 'in'.

1 p.86 The sixth colour of the rainbow.

2 p.112 Global network linking servers and personal computers.

3 p.152 Imperial measurement equal to just over 2.5cm.

4 p.44 A quantity larger than any conceivable number.

5 p.106 The period of history characterised by the introduction of mass production.

6 p.86 Radiation given off by hot objects, just outside the visible spectrum.

7 p.140 To give vaccine for immunity against disease.

8 p.66 The world's third-largest producer and consumer of coal.

9 p.142 A hormone injected to control diabetes.

10 p.118 In chemistry, the term for all non-carbon compounds and some simple carbon compounds.

Medical solutions
Ten medical questions.

11 p.142 Which came first, the first kidney transplant or the first heart transplant?

12 p.140 Which medical instrument, today Y-shaped, was invented by René Laënnec in 1816?

13 p.140 Until recently, doctors were required to take an oath on the name of which ancient Greek physician?

14 p.140 A, B and O are three of the four main blood groups. What is the fourth?

15 p.142 Which mould links Alexander Fleming with Camembert cheese?

16 p.142 How is the mild anaesthetic nitrous oxide better known?

17 p.140 Which short word follows cow, small and chicken to give the names of three diseases?

18 p.140 To what part of the body does the word cardiac refer?

19 p.142 Florence Nightingale is credited with starting the modern nursing profession. In which European city was she born?

20 p.142 What type of flower yields the extracts from which morphine is made?

Pot luck
A mixed selection of teasers.

21 p.88 The HST was named after the man who discovered that the Universe is expanding. What is the HST?

22 p.48 How many coloured squares are there on a Rubik's cube?

23 p.56 In which world-famous shop was the first escalator in Britain installed?

24 p.102 How many pieces of A4 paper can be cut from an A1 sheet?

25 p.56 What part of a building can be saddleback, hip, curb or gable?

26 p.68 Which US nuclear power plant came within half an hour of meltdown in 1979?

27 p.88 What was Shoemaker-Levy, which crashed into Jupiter in 1992?

28 p.118 Why would someone put H_2O_2 on their hair?

29 p.148 Max Planck, Niels Bohr and Albert Einstein all won which Nobel prize?

30 p.74 Which famous motorcycle manufacturer's first model had a tomato can for a carburettor?

Mixed-up maths
Use the clues to unravel the anagrams next to them.

31 p.48 READITEM The width of a circle.

32 p.46 INCAFORT Part of a whole number.

33 p.44 MEDICAL Number system that developed from counting on fingers.

34 p.50 GEARAVE Not the median or mode, but the mean.

35 p.46 CAPEREGENT Proportion represented as parts of 100.

36 p.40 CHIMETRAIT The art of calculating.

37 p.48 TROPHYSAGA Ancient Greek who saw the world in triangles.

38 p.50 BABYOILTRIP Field of maths concerned with the likelihood of events happening.

39 p.44 RIPME BURNME Can be divided only by one and itself.

40 p.48 EARNCIDERPULP A line at right angles to a given plane or surface.

**For answers and more facts go to the page given below each question number.
For quick answers to complete quizzes 0 to 6 go to page 154.**

On the road
Automobile themed puzzles.

41 p.74 — Who returned to driving an Aston Martin on screen in 2002 after having driven a BMW through the 1990s?

42 p.124 — By what name is the bulbous, white rubber character Bibendum better known?

43 p.106 — Motown music originated in Detroit. What is Motown short for?

44 p.74 — What type of car featured in the film *The Love Bug*?

45 p.74 — Which motorbike manufacturer has ranges called Goldwing and Silverwing?

46 p.74 — In which decade did the first-ever automobile race take place?

47 p.74 — What does the CV in Citroën 2CV stand for?

48 p.74 — Which disastrous motor car did Henry Ford name after his son?

49 p.74 — Which US car manufacturer once made revolutionary vacuum cleaners?

50 p.74 — Where in a new car would you expect to find a catalytic convertor?

Great works
Questions on scientific literature.

51 p.54 — Whose Special Theory of Relativity was published in 1905?

52 p.134 — Which complex molecule was the reason for James Watson's biography *The Double Helix*?

53 p.144 — In which Mary Shelley novel was electricity used to bring dead matter to life?

54 p.144 — Which modern physicist wrote *A Brief History of Time* and *The Universe in a Nutshell*?

55 p.144 — Which zoologist and television presenter wrote *The Naked Ape*?

56 p.144 — Who expanded on the work of Darwin with *The Blind Watchmaker* and *The Selfish Gene*?

57 p.144 — Which dramatised version of a novel caused widespread panic when read on the radio by Orson Welles in 1938?

58 p.130 — James Lovelock's unifying theory of life on Earth was named after which Greek deity?

59 p.136 — Which landmark book by Rachel Carson exposed the dangers of the pesticide DDT?

60 p.144 — Who wrote *Utopia* in 1516, considered by many to have been the first ever science-fiction novel?

'S' is for science
Select the correct option from the four possible answers.

61 p.74 — Which of these first appeared as part of an automobile?

A Spark plug	B Safety belt
C Speedometer	D Supercharger

62 p.42 — Which weapon or tool is acknowledged to be the oldest?

A Spear	B Sickle
C Sword	D Saw

63 p.104 — In what scientific process might a laser be used to determine the chemical composition of a substance?

A Separation	B Syncopation
C Spectroscopy	D Synergy

64 p.90 — What does the 's' in sonar stand for?

A Surface	B System
C Sea	D Sound

65 p.114 — In what process do heated solids turn directly to gas?

A Sublimation	B Saturation
C Solification	D Stupefaction

66 p.116 — Which of these is not an element in the Periodic Table?

A Samarium	B Stibnite
C Silicon	D Scandium

67 p.98 — What do the letters SLR mean when applied to a camera?

A Single-lens reflex	B Self-loading reel
C Standard light ratio	D Scanning long-range

68 p.118 — Which of these has the highest pH scale value?

A Soapy water	B Sea water
C Saliva	D Stomach acid

69 p.64 — Which British inventor produced light bulbs at the same time as Edison, or possibly even before him?

A Swan	B Spencer
C Stanley	D Sturgeon

70 p.80 — What was Earth's first geosynchronous satellite called?

A Sputnik	B Soyuz
C Syncom	D Sparrow

First and last
All solutions begin and end with the same letter.

71 p.46 A branch of mathematics that uses letters to represent numbers.

72 p.98 The photographic company responsible for the 'Brownie' box camera.

73 p.116 An inert, noble gas used in tubes to make colourful signs.

74 p.100 The point on the needle of a record player.

75 p.48 The property of a circle found by multiplying π (pi) by the square of the radius.

76 p.64 A light emitting device refined by Thomas Edison.

77 p.104 The company that introduced the photocopier and invented the laser printer.

78 p.64 Food cooler developed by Karl von Linde.

79 p.84 An explosive device.

80 p.128 A synthetic fibre used in tights and stockings.

Pot luck
A mixed selection of teasers.

81 p.108 Which company has become famous for its 'Pentium' chips?

82 p.124 What colour was woad, widely used as a dye until the 1800s?

83 p.82 Butterfly, flick and stiletto are all what type of weapon?

84 p.70 What did Don Quixote do battle with, convinced they were giants?

85 p.114 Are atoms made of molecules, or molecules made of atoms?

86 p.124 Which common transparent material has both solid and liquid properties at room temperature?

87 p.84 What seven-letter word is applied to aircraft that cannot be detected by radar?

88 p.148 Which Russian, best-known for his drooling dogs, was awarded the Nobel prize for medicine in 1904?

89 p.124 What fibrous mineral that causes lung cancer can be usefully woven into fireproof clothing?

90 p.114 Complete this children's film title: *Jimmy _____, Boy Genius*?

Number crunchers
A series of mathematical teasers.

91 p.74 How many wheels are there on a Segway?

92 p.44 How many factors does a prime number have?

93 p.46 Which is the denominator of a fraction – the upper or lower number?

94 p.44 Which word mainly used today to mean 'many', originally meant 10 000?

95 p.46 What branch of mathematics derives its name from the Latin for pebble?

96 p.44 How many millions are there in a billion, when referring to money?

97 p.44 Which number system, or base, do electronic calculators use?

98 p.144 What is the answer to life, the Universe and everything, according to the computer Deep Thought?

99 p.46 Nought point three (0.3) recurring is the same as what proper fraction?

100 p.90 How many degrees of latitude are there between the North Pole and the Equator?

Reaching for the stars
Ten questions on outer space.

101 p.80 What first did Valentina Tereshkova achieve on June 16, 1963?

102 p.80 Which near-disastrous Moon mission was made into a blockbuster film starring Tom Hanks?

103 p.54 What event in 1919 helped Einstein to prove his theory that starlight bends near the Sun?

104 p.88 Apart from Earth, which planet did not appear in Gustav Holst's orchestral suite *The Planets*?

105 p.80 What nationality was the first woman in space?

106 p.68 Which natural nuclear reactor is the best known Yellow Dwarf in the G2 spectrum?

107 p.88 SETI is an international project searching for what?

108 p.80 Which planet is denoted by the biological symbol for male?

109 p.80 Which film starring Sigourney Weaver was promoted with the line 'In space, no one can hear you scream'?

110 p.80 Which US Senator returned to space in 1998 at the age of 77?

**For answers and more facts go to the page given below each question number.
For quick answers to complete quizzes 7 to 13 go to page 154.**

Up and over
A quiz on skyscrapers, bridges and other buildings.

111 p.56 Which English building has housed a zoo, an observatory, a mint and a prison?

112 p.56 What shape is the 36-storey Las Vegas Luxor Hotel?

113 p.58 What type of bridge carries water over a valley?

114 p.58 In which century was the first-ever iron bridge built?

115 p.58 In 1928, what was then the world's longest single-span bridge was opened in which English city?

116 p.58 What 'c' describes the structure of the Forth Rail Bridge and the Quebec Bridge?

117 p.56 Which landmark, widely disliked when built in 1884, was referred to as 'the tragic lamppost'?

118 p.56 What was the world's tallest building for more than 3000 years?

119 p.58 Which two countries are linked by the Öresund Bridge?

120 p.56 In which city was the first steel-framed skyscraper built?

Come rain or shine
Sunshine, showers and the science behind them.

121 p.130 Photosynthesis in plants is driven by sunlight. What is the green pigment that facilitates photosynthesis called?

122 p.130 Which atmospheric gas is a by-product of photosynthesis?

123 p.130 Which element, made useful to plants by lightning, is transferred to their roots by rain?

124 p.68 What does a smiling Sun have in common with an inverted trident in a circle?

125 p.52 What type of clock only works during daylight?

126 p.66 Which fuel used in power stations is the main cause of acid rain?

127 p.70 In which US state is the world's largest solar power plant?

128 p.88 Which telescope was created to observe the Universe without interference from weather or other atmospheric effects?

129 p.70 What development in 1981 meant that maths students no longer had to buy batteries?

130 p.116 Which element was discovered on the Sun 19 years before it was found on Earth?

Fertile minds
Select the correct option from the four possible answers.

131 p.126 What invention by Henry Brearley changed cutlery forever?

A The fish knife	B Serrated cutting edges
C Bone handles	D Stainless steel

132 p.102 What spread the word about its inventor Gutenberg?

A The printing press	B The radio
C The loudspeaker	D The mobile phone

133 p.148 What was invented by Alfred Nobel, of peace-prize fame?

A Dynamite	B The delayed fuse
C Kevlar	D Solar panels

134 p.84 Who designed the original AK47?

A Alexei Kievan	B Vladimir Stolypin
C Mikhail Kalashnikov	D Aleksander Kerensky

135 p.64 Why does Las Vegas owe a debt to Georges Claude?

A He invented slot machines	B He invented roulette
C He discovered electricity	D He invented neon lights

136 p.64 Who invented the electric guitar?

A Charlie Christian	B Adolph Rickenbacker
C James Fender	D Jackie Wilson

137 p.112 Whose work led to the creation of the World Wide Web?

A Bill Gates	B Tim Berners-Lee
C Vint Cerf	D Alan Turing

138 p.108 Who began work on a 'difference engine', which was not completed until 120 years after his death?

A Robert Stephenson	B Karl Benz
C Charles Babbage	D Leonardo da Vinci

139 p.124 William Perkin's attempts to synthesise quinine led to what?

A First artificial sweetener	B First plastic
C First synthetic dye	D First adhesive tape

140 p.46 Who first used the decimal point in a book?

A Christoph Rudolf	B John Napier
C Jacob Bernoulli	D Henry Briggs

True or false?
Decide whether the given facts are correct or not.

141 p.70 The world's biggest dam is on the river Nile.

142 p.58 You can see the Great Wall of China from the Moon with the naked eye.

143 p.138 The sandwich is named after the fourth Earl of Sandwich.

144 p.60 The crack of a whip is actually a sonic boom as it breaks the sound barrier.

145 p.82 The halberd was a weapon used by medieval infantrymen.

146 p.102 Indian ink originated in India.

147 p.50 Florence Nightingale invented the pie chart.

148 p.142 Hypodermic means under the skin.

149 p.138 Bird's Eye Foods was founded by Clarence Birdseye.

150 p.78 The Wright Brothers' aeroplane was called the Kitty Hawk.

Power struggles
All of these anagrams relate to power generation.

151 p.70 WILLMIND Harnesses a natural energy source to generate power.

152 p.68 NEWTMOLD Worst-case nuclear scenario – what happened at Chernobyl.

153 p.68 CHINA ACNERIOT One thing leads to another in this nuclear term.

154 p.70 OGREHAMLET The term for energy produced from the Earth's natural heat.

155 p.68 INOUTLUMP The second most commonly used nuclear fuel.

156 p.70 MONDAY Device for generating electricity, sometimes by using a bicycle wheel.

157 p.66 RUINBETS Engines driven by the flow of water, steam or gas.

158 p.68 TIARAVOICED SWEAT A dangerous by-product of nuclear power.

159 p.114 CANRULE IFSINSO The process of splitting the atom.

160 p.66 NOLOGIC ORSTEW Chimney-like structures that emit nothing but steam.

Pot luck
A mixed selection of teasers.

161 p.132 What type of animal cells are thrombocytes and leucocytes?

162 p.54 What 'f' explains why flat-sided objects are more difficult to move than spherical ones?

163 p.88 When a lunar eclipse occurs, which body's shadow do we see projected on the Moon?

164 p.134 By what natural circumstance can two people end up with the same DNA?

165 p.64 Which is not found inside lightbulbs: argon, krypton or oxygen?

166 p.110 Which vowel is not on the top row of letters on a standard computer keyboard?

167 p.92 The 'circle of light' is a fibre-optic communications cable surrounding which continent?

168 p.146 The technology of extreme miniaturisation is known as what?

169 p.98 Fox Talbot's *The Pencil of Nature* was the first-ever book of what?

170 p.120 Uranium 238 has three more what than Uranium 235?

Magic formulae
Questions on theories, laws and formulae.

171 p.48 What does the formula π (pi) x diameter determine?

172 p.48 Whose theorum enables the length of the hypotenuse to be calculated when all that is known are the lengths of the adjacent and opposite sides?

173 p.130 Glucose plus oxygen turns to water, energy and what?

174 p.150 Whose law relates volume and pressure in gases?

175 p.48 What property of a parallelogram can be found by multiplying the length of its base by its height?

176 p.54 Who theorised that an object's mass is always constant but that its weight changes as it accelerates?

177 p.62 In electronics, voltage equals current multiplied by what?

178 p.48 If one of the internal angles of a rhombus is 75° what are the other three?

179 p.86 What property of electromagnetic waves can be determined by dividing the speed of light by the waves' frequency?

180 p.150 Whose law states that equal volumes of all gases at the same temperature contain the same number of molecules?

For answers and more facts go to the page given below each question number.
For quick answers to complete quizzes 14 to 18 go to page 154.

QUIZ
18
WARM UP

Structural identity
Can you identify these structures? Match them to the names below.

Akashi-Kaikyo Bridge

Arecibo telescope

CN Tower

Eiffel Tower

Empire State Building

Iron Bridge

London Eye

Millennium Dome

Petronas Towers

Sydney Harbour Bridge

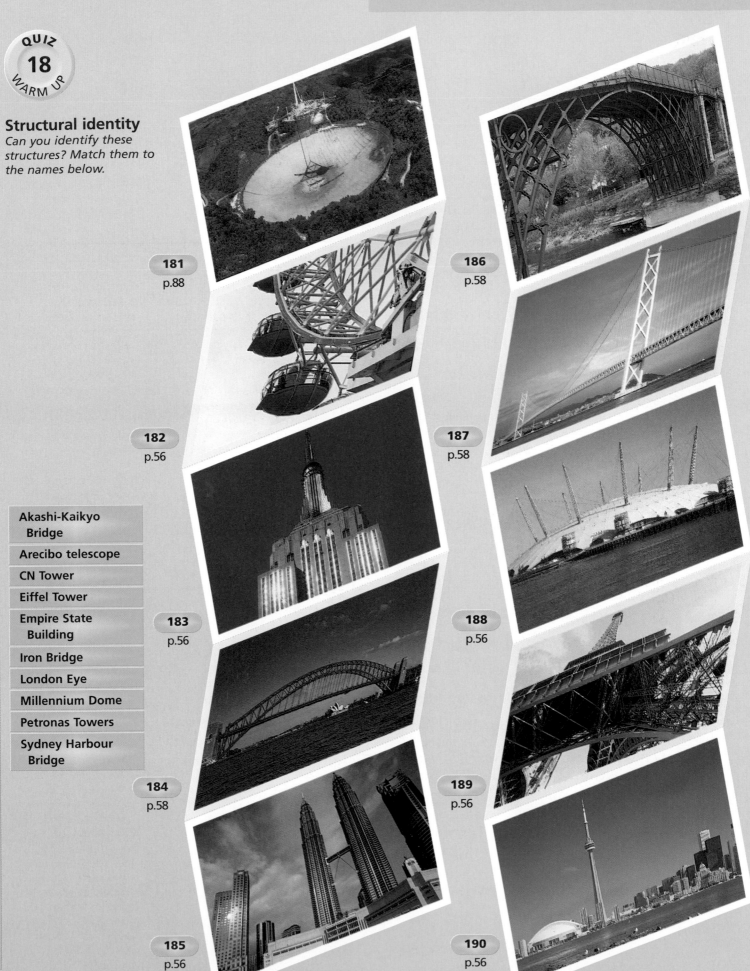

181 p.88

182 p.56

183 p.56

184 p.58

185 p.56

186 p.58

187 p.58

188 p.56

189 p.56

190 p.56

Fighting talk
Questions on weapons or linked with battle and war.

191 p.84 To which Caribbean island did the Soviet Union ship nuclear missiles in 1962?

192 p.82 Which medieval weapon shares its name with a defensive spray and a spice?

193 p.118 In which country was gunpowder invented?

194 p.86 What does the phrase 'Richard of York gave battle in vain' help you to remember?

195 p.84 What type of weapon was patented by Richard Gatling in 1862?

196 p.108 Which country used the Enigma Machine to generate codes in the Second World War?

197 p.84 Which First World War German fighter pilot is often mentioned by Snoopy in *Peanuts*?

198 p.82 The surname Fletcher implies ancestors who made what type of weaponry?

199 p.82 Cutlass and claymore were both types of what?

200 p.84 What missiles used by Iraq in 1991 originated in the Soviet Union in the mid-1960s?

Man-machine
A quiz mainly about robots.

201 p.106 Which 1984 movie starred Arnold Schwarzenegger as a murderous cyborg?

202 p.142 What was the Tin Man searching for in *The Wizard of Oz*?

203 p.106 What is the name of the cowardly robot on the starship *Red Dwarf*?

204 p.110 Which 2001 Spielberg film was promoted with the line 'His love is real. But he is not'?

205 p.106 Sony's robot Aibo is built in the shape of what animal?

206 p.106 What was the name of Dr Who's robot companion?

207 p.106 Who wrote 'Runaround' in 1942, a story that proposed three famous laws of robot behaviour?

208 p.106 Which law-enforcing cyborg was programmed to 'protect the innocent' in one of 1987's biggest-grossing films?

209 p.106 Which 'paranoid android' travelled with Arthur Dent in the *Hitchhiker's Guide to the Galaxy* novels?

210 p.142 What was Barney Clark the first patient to receive, in 1982?

Growing concerns
Ten questions on cultivating life.

211 p.134 What is the name of the sheep cloned at the Roslin Institute in 1997?

212 p.136 Which famous English public school is also a device for tilling soil?

213 p.136 What 'h' is an animal or plant produced by crossing two different species?

214 p.136 What 'h' is the technique of growing plants in nutrient solutions instead of soil?

215 p.136 What material is added to living cells by genetic engineering?

216 p.136 How must food be produced in order to qualify for certification by the Soil Association?

217 p.136 Which notorious pesticide led to the collapse of the USA's bald eagle population in the 1960s and 70s?

218 p.136 Which invention of 1884 revolutionised the process of harvesting crops?

219 p.136 What type of farm is owned by The Who's Roger Daltrey at Iwerne Springs in Dorset?

220 p.134 How did a weed called thale cress make history in the year 2000?

Sound reasoning
A quiz on sound recording and reproduction.

221 p.100 Which jukebox manufacturer first made its name producing pipe organs?

222 p.100 Who famously recorded the nursery rhyme 'Mary Had a Little Lamb' on a tinfoil-covered cylinder in 1877?

223 p.96 What was revolutionary about the Freeplay radio, invented by Trevor Bayliss in 1991?

224 p.100 What three-dimensional sound reproduction effect takes its name from the Greek word for 'solid'?

225 p.100 Which music storage format was invented by Sony's president to store Beethoven's 9th Symphony?

226 p.100 Which recent addition to the English language means 'empty orchestra' in Japanese?

227 p.100 What is the name of the dog listening to a gramophone on the HMV logo?

228 p.100 What two words link the author Nick Hornby to sound reproduction?

229 p.100 How many grooves are there on a typical vinyl record?

230 p.100 Which of the notes above middle C is used to define concert pitch, usually 440 Hz?

For answers and more facts go to the page given below each question number.
For quick answers to complete quizzes 19 to 25 go to pages 154 and 155.

Making waves
Ten questions on waves.

231 p.72 Which nautical movie was the first film to earn over US$1 billion at the box office?

232 p.70 Which heavenly body is the driving force behind tidal power?

233 p.72 A boat with two hulls is called a catamaran. What is a boat with three hulls called?

234 p.86 What type of waves cause sunburn and skin cancer?

235 p.86 What type of waves are used to carry the messages picked up by most mobile phones?

236 p.86 Wavelength is the distance between successive wave crests. What term describes the number of wavelengths per second?

237 p.86 What colour of light in a rainbow has the longest wavelength?

238 p.140 Which waves revolutionised medicine when they were first used in 1895?

239 p.86 Which electromagnetic waves are often formed during nuclear reactions?

240 p.88 What type of waves are detected by the Jodrell Bank telescope?

Pot luck
A mixed selection of teasers.

241 p.140 Wilhelm Röntgen took the world's first X-ray photograph in 1895. What was the subject?

242 p.112 What term was coined by author William Gibson in 1981 in a short story about an electronic network?

243 p.50 What does '50-1 bar' mean at the bottom of a table of racing odds?

244 p.98 What was the first-ever feature-length animated film?

245 p.54 Which scientist and MP spoke only once in England's Parliament, to ask for a window to be opened?

246 p.88 Which 17th-century Italian scientist was released from purgatory by the Pope in 1992?

247 p.148 Which spoof prizes are awarded each year by the magazine *Annals of Improbable Research*?

248 p.112 *Wax: Or The Discovery of Television Among the Bees* became the first-ever what in 1993?

249 p.102 In printing terminology, what are referred to as the verso and the recto?

250 p.118 Dry ice, used in smoke machines, is the frozen form of which common chemical compound?

Words and pictures
Select the correct option from the four possible answers.

251 p.112 Which of these is a term for unwanted 'junk' email?

A Ham	B Spam
C Jam	D Yam

252 p.100 How was the movie-going experience improved in 1941?

A Stereo sound in cinemas	B Ticketed seating
C Invention of popcorn	D Wearing hats banned

253 p.124 What was medieval parchment made from?

A Papyrus leaves	B Dried animal skins
C Willow bark	D Straw and glue

254 p.100 When did the first domestic video camera go on sale?

A 1964	B 1972
C 1980	D 1989

255 p.102 What do the abbreviations CMYK represent?

A Colours	B Sounds
C Numbers	D Chemical elements

256 p.108 Which of the following encodes visual data for a computer?

A Scanner	B Zip disc
C Photocopier	D Server

257 p.102 Who produced the first-ever illustrated printed book?

A Albrecht Pfister	B William Ged
C George Harper	D Joseph Collins

258 p.102 When was the first desktop publishing software introduced?

A 1975	B 1984
C 1992	D 1997

259 p.102 Which of the following is not a sans serif font?

A Avant Garde	B Helvetica
C Arial	D Palatino

260 p.102 Who invented full-colour printing in 1719?

A Alois Senefelder	B Johann Fust
C Peter Schöffer	D Jakob Le Blon

True or false?
Decide whether the given facts are correct or not.

261 p.114 O_3 is the chemical formula for ozone.

262 p.116   The lead in pencils is actually carbon.

263 p.40   Taxonomy is the science of taxation.

264 p.150   Avogadro's Law, Boyle's Law and Charles's Law all concern gases.

265 p.58   The Channel Tunnel is the world's longest railway tunnel.

266 p.132   All living things are made of more than one cell.

267 p.104   Laser is an abbreviation of 'light amplification by stimulated emission of radiation'.

268 p.60   A calorie is the energy required to heat a litre of water by 1°C (one degree Celsius).

269 p.152   A metric tonne is heavier than an Imperial ton.

270 p.104   The US has a working laser for shooting down missiles.

Anagrammed arms
Use the clues to unravel the anagrams next to them.

271 p.82   WORCSOBS Powerful weapon that fired bolts or quarrels.

272 p.82   BEARS Curved sword used by cavalrymen.

273 p.82   TUSKME Predecessor of the rifle.

274 p.84   MINUSBEACH GNU Rapid-firing weapon developed between the two World Wars.

275 p.84   RODPOET Self-propelled missile carried by submarines.

276 p.84   HEROFELTWARM Weapon that fires a jet of combustible liquid.

277 p.84   DRIEDSINEW Air-to-air missile named after a desert rattlesnake.

278 p.84   TRAMS SAWOPEN Modern arms designed to hit targets with pinpoint accuracy.

279 p.82   WITHZERO Short cannon invented by the Dutch in 1740.

280 p.82   TETHERCUB Medieval siege engine for throwing rocks and other large missiles.

Getting the message
A quiz on multi-media communications.

281 p.94   Which country is the home of mobile phone giant Nokia?

282 p.94   Which number key on a telephone keypad bears the letters PQRS?

283 p.94   In text-message language, which American saying does HAND represent?

284 p.94   Text messaging is also known as SMS. What does SMS stand for?

285 p.92   In which country do 'School of the Air' teachers mark pupils' homework by fax?

286 p.92   What is the word fax short for?

287 p.94   What flirtatious question would you be asking if you text someone the message WYGOWM?

288 p.92   Who, in 1876, famously called for his assistant with the words 'Mr Watson, come here, I want you'?

289 p.94   WAP phones give limited access to the Internet. What three words do the letters WAP stand for?

290 p.92   In which century was the first telephone exchange opened?

Screen test
Television and film teasers.

291 p.96   What three colours are used to generate all of the shades on a colour TV screen?

292 p.96   Which organisation made its first television broadcasts from Alexandra Palace?

293 p.96   What was the name of the world's first television satellite?

294 p.96   In which decade did regular colour television transmission start in the UK?

295 p.100   Which system designed by Sony was beaten by VHS in the 'war of the video formats'?

296 p.96   What tiny particles are fired at a TV screen to make it light up?

297 p.100   In the 1927 film *The Jazz Singer*, who delivered the now famous line 'You ain't heard nothing yet!'?

298 p.96   What type of television transmission would be made from an 'OB' unit?

299 p.96   What 'friendly' three-letter acronym describes the UK's television transmission standard?

300 p.96   What is the claim to fame of office boy William Taynton?

**For answers and more facts go to the page given below each question number.
For quick answers to complete quizzes 26 to 32 go to page 155.**

Pot luck
A mixed selection of teasers.

301 p.116 Which element, accounting for more than three-quarters of all matter, is the most common in the Universe?

302 p.68 Which country is the world's largest producer of nuclear waste?

303 p.80 Quartz watches were originally designed for use where?

304 p.92 What emergency signalling system was replaced in 1999 by the Global Maritime Distress and Safety System?

305 p.134 Which 1997 film starring Uma Thurman and Ethan Hawke raised the issue of genetic make-up dictating social status?

306 p.150 What sort of bodies do Kepler's Laws concern?

307 p.64 Who said 'Genius is one per cent inspiration and 99 per cent perspiration'?

308 p.122 To the nearest five gallons, how many Imperial gallons of oil are there in a standard barrel?

309 p.104 Lasers work with visible light. What did the laser's predecessor, the maser, use?

310 p.116 What 'i' means different forms of the same chemical element?

Four of a kind
Give the term that best describes all four examples.

311 p.108 Daisywheel, dot-matrix, inkjet, laser.

312 p.148 Physics, chemistry, literature, peace.

313 p.134 Adenine, cytosine, guanine, thymine.

314 p.60 Potential, kinetic, chemical, nuclear.

315 p.72 Caravel, ketch, barque, galleon.

316 p.132 Glucose, fructose, ribose, sucrose.

317 p.64 DC, universal, linear induction, linear synchronous.

318 p.102 Letterpress, gravure, offset lithography, photolithography.

319 p.132 Mitochondrion, Golgi apparatus, lysosome, endoplasmic reticulum.

320 p.146 Meme, CAVE, ARGOS, PHANToM.

Matters of life and death
Select the correct option from the four possible answers.

321 p.130 According to Charles Darwin, the process by which species change over time is evolution through what?

A The eye of a needle	B Hell and high water
C Natural selection	D Creationism

322 p.40 What is the scientific study of animals called?

A Zoology	B Psychology
C Faunology	D Astrology

323 p.132 Which of these is the largest?

A Red blood cell	B White blood cell
C Egg cell	D Sperm cell

324 p.130 Which word means the release of water vapour by plants?

A Transpiration	B Perspiration
C Distillation	D Precipitation

325 p.130 In biology, the conversion of nitrogen into nitrates by bacteria is called what?

A Restoring	B Repairing
C Mending	D Fixing

326 p.132 Which of the following have chloroplasts?

A Animals	B Plants
C Fungi	D Bacteria

327 p.134 How many chromosomes do most human beings have?

A 28	B 46
C 64	D 92

328 p.134 What type of cell division creates sex cells?

A Mitosis	B Meiosis
C Microprocess	D Myxomatosis

329 p.134 Approximately what proportion of human genes are also found in mice?

A 10 per cent	B 30 per cent
C 60 per cent	D 90 per cent

330 p.132 What is the main function of ribosomes in a cell?

A Protein synthesis	B Breaking down fats
C Transferring DNA	D Maintaining cell structure

18 QUIZ QUESTIONS 331 to 400

 QUIZ 33 WARM UP

 QUIZ 34 IN YOUR STRIDE

 QUIZ 35 IN YOUR STRIDE

 QUIZ 36 ALL COMERS

Changing the menu
Answer the following questions on foodstuffs.

331 p.138 In which US state did brothers Maurice and Richard McDonald open their 'assembly-line' hamburger restaurant in 1940?

332 p.136 Which 'labyrinthine' crop was introduced to Europe in the 1500s?

333 p.138 Which French chemist and microbiologist introduced the technique of gently heating food to partially sterilise it?

334 p.138 Still popular today, which food product was the first to be sold in aerosol form?

335 p.138 Which sweetener, invented in 1879, is about 300 times sweeter than sugar?

336 p.138 Whose secret recipe is advertised as containing '11 herbs and spices'?

337 p.142 A, B, D, E, K: which letter is missing?

338 p.138 In which century was margarine invented?

339 p.138 What was the first natural flavouring to have an artificial version made of it?

340 p.136 What dubious distinction was bestowed on tomato paste in 1996?

Pot luck
A mixed selection of teasers.

341 p.134 DNA stands for deoxyribonucleic what?

342 p.42 What secret does King Louie try to learn from Mowgli in Disney's *The Jungle Book*?

343 p.74 What type of engine requires oil to be mixed with petrol to work?

344 p.144 'Any sufficiently advanced technology is indistinguishable from magic' is the third law of which science-fiction writer?

345 p.58 What 'c' is the curve on a road surface, built in to encourage drainage?

346 p.152 A temperature of 50° Fahrenheit is equivalent to how many degrees Celsius?

347 p.84 The atomic bomb dropped on Nagasaki was called Fat Man. What was the bomb dropped over Hiroshima called?

348 p.56 What people-carrying device, first put into service in a New York department store in 1857, was one key factor in making skyscrapers possible?

349 p.90 What is the most common application of radar technology by modern police forces?

350 p.152 What unit of frequency, named after a 19th-century German physicist, is defined as one cycle per second?

All about image
Photography and camera-related questions.

351 p.98 What type of lens, often used by the paparazzi, has a focal length of over 85 mm?

352 p.98 Which instant photo company, founded by Edwin Land, went bust in October 2001?

353 p.98 What type of device is normally attached to a camera's 'hot shoe'?

354 p.98 What is the name of the toothed wheel that pulls film through a camera?

355 p.98 Which British photographer inspired the classic 1966 film *Blow-Up*?

356 p.140 Which wave-based technology lets doctors take pictures of babies before they are born?

357 p.98 What type of cameras can be judged in quality by their megapixel capacity?

358 p.98 What type of animal did Eadweard Muybridge photograph to settle a bet about how it ran?

359 p.98 What on a camera is measured in 'f-stops'?

360 p.98 What name is given to a dark room where images of outside objects are projected in through a small hole?

Travel and transport
Ten questions on planes, trains and automobiles.

361 p.74 What type of vehicle was a 'penny farthing'?

362 p.76 In which country will you find TGVs, the world's fastest trains?

363 p.76 What 'one-track' transport system links the US Capitol building with the Congressional offices?

364 p.78 Why are modern air ships filled with helium rather than hydrogen?

365 p.72 What type of ship is HMS *Ark Royal*?

366 p.72 What did Christopher Cockerell develop after experimenting with a vacuum cleaner and a coffee tin?

367 p.74 Ethylene glycol is very useful in winter. What is it better known as?

368 p.80 What kind of spacecraft made its first flight in April 1981?

369 p.78 What sea-going vessels were built by the Boeing Company before 1940?

370 p.72 On what might you find a bowsprit, mizzen and spanker?

For answers and more facts go to the page given below each question number.
For quick answers to complete quizzes 33 to 39 go to page 155.

Forces of nature
A quiz on energy and physical forces.

371 p.62 What type of electricity, despite its name, can make small, light objects move?

372 p.70 Where would you find 'nodding ducks' generating electricity?

373 p.54 Which of Newton's Laws of Motion is most associated with the falling apple story?

374 p.54 How many times stronger is the Earth's gravitational pull than that of the Moon?

375 p.54 Newton's third Law of Motion states that every action has an equal and opposite what?

376 p.80 Where was David Scott when he simultaneously dropped a feather and a hammer to prove a famous gravity theory?

377 p.62 If you stroke an iron bar with a lodestone then let it swing freely, what happens?

378 p.86 Four 'fundamental forces of nature' direct how elementary particles interact: the weak nuclear force, the strong nuclear force, gravity – and what?

379 p.120 The half-life of strontium-90 is 28 years. What mass will 1 kg of strontium-90 decay to after 56 years?

380 p.120 Which unit of radioactivity is named after the French scientist who discovered that uranium was radioactive?

First and last
All solutions begin and end with the same letter.

381 p.110 A device for modulation and demodulation connecting a computer to a telephone line.

382 p.116 A metal that burns with a bright, white light.

383 p.90 A word coined from an abbreviation of 'radio detecting and ranging'.

384 p.114 To change from liquid into vapour.

385 p.120 The Greek letter given to a radioactive particle of two protons and two neutrons.

386 p.116 The seventh element in the Periodic Table.

387 p.122 The toughest and most valuable form of carbon.

388 p.92 A device that boosts and re-transmits signals on a long-distance telephone cable.

389 p.54 A physicist famous for his three Laws of Motion.

390 p.54 The scientific term for the quantity of motion of a moving body.

Record breakers
Select the correct option from the four possible answers.

391 p.78 What is the world's fastest commercial jet airliner?

| A Concorde | B Boeing 747 |
| C Airbus A320 | D de Havilland Comet |

392 p.76 Which of these is the longest?

| A World's longest ship | B World's longest car |
| C World's longest aeroplane | D World's longest train |

393 p.56 In which European city is the world's largest dome?

| A Rome | B Vienna |
| C Berlin | D London |

394 p.58 The world's longest suspension bridge links islands in which country?

| A Canada | B Japan |
| C Indonesia | D Denmark |

395 p.56 What is the world's tallest structure?

| A CN Tower | B Petronas Towers |
| C Sears Roebuck Tower | D Empire State Building |

396 p.148 Who was the youngest person ever to win a Nobel prize?

| A William Bragg | B Max Planck |
| C Georges Charpak | D Albert Einstein |

397 p.74 What is the best-selling car of all time?

| A Volkswagen Golf | B Model T Ford |
| C Toyota Corolla | D Vauxhall Cavalier |

398 p.58 What is the world's longest ship canal system?

| A Suez Canal | B St Lawrence Seaway |
| C Erie Canal | D Panama Canal |

399 p.58 Where is the world's longest fence?

| A Russia | B Australia |
| C Argentina | D Kenya |

400 p.58 Which river is blocked by the world's tallest dam?

| A Colorado, USA | B Yangtze, China |
| C Araguaia, Brazil | D Vakhsh, Tajikistan |

Pot luck
A mixed selection of teasers.

401 p.110 — What does the 'm' stand for in the computer terms ROM and RAM?

402 p.46 — Which is larger: two metres square or two square metres?

403 p.120 — Uranium eventually decays into which old-fashioned roofing material?

404 p.126 — What must be added to iron in order to make steel?

405 p.80 — What is the difference between a cosmonaut and an astronaut?

406 p.110 — British designer Jonathan Ive has won acclaim for his work with which computer company?

407 p.56 — What floor do British buildings have that US buildings do not?

408 p.126 — What substances are joined together using solder?

409 p.134 — Who have Y chromosomes, men or women?

410 p.144 — In *Around the Moon*, what did Jules Verne correctly predict would happen to people in space?

Modes of travel
Questions on planes, trains and automobiles.

411 p.78 — Which supersonic jet stretches by up to 25 cm (10 in) during flight?

412 p.74 — Which US state has the world's tightest restrictions on exhaust fumes?

413 p.106 — Who said about his cars 'you can have it in any colour as long as it's black'?

414 p.78 — What unpowered aircraft gain height by riding on thermals?

415 p.76 — In the books by Rev. W. Awdry, how is the character Sir Topham Hatt better known?

416 p.76 — Which national rail company has over 1 600 000 staff and is the world's largest civilian employer?

417 p.76 — What pulled the carriages on the earliest railways?

418 p.78 — Which actor said 'I am serious, and don't call me Shirley!' in the movie *Airplane!*?

419 p.52 — What can be calculated by dividing the distance travelled by the time spent travelling?

420 p.84 — What was the name of the aeroplane that dropped the atomic bomb over Hiroshima?

All bar ones
All the names start with the word 'bar'.

421 p.84 — The tube on a gun through which bullets are fired.

422 p.52 — An instrument for measuring changes in atmospheric pressure.

423 p.104 — A series of lines and numbers used to identify a product.

424 p.72 — A long vessel for transporting goods over water, especially inland waterways.

425 p.84 — A heavy bombardment by artillery.

426 p.82 — A fortification at the gate of a medieval town.

427 p.142 — A group of drugs with sedative and anaesthetic properties.

428 p.142 — The South African surgeon who carried out the world's first heart transplant.

429 p.116 — A soft, white, metallic element; number 56 in the periodic table.

430 p.148 — The joint inventor of the transistor and superconductivity pioneer, twice awarded the Nobel prize for physics.

Muddled monikers
Use the clues to unravel the anagrams next to them.

431 p.62 — FINNLARK American statesman who invented the lightning conductor.

432 p.72 — OUTSAUCE Famous French diver who invented the aqualung.

433 p.62 — GINLAVA Physiologist who used frogs' legs to demonstrate that nerve action involved electricity.

434 p.46 — RAFTME French mathematician who founded modern number theory.

435 p.104 — IMAMAN American physicist who built the first laser.

436 p.124 — NOTKIPLING Company, named after a major shareholder, that invented float glass.

437 p.140 — USALEVIS Anatomist whose medical drawings of the human body were the first ever published.

438 p.150 — BEGREENISH German physicist who gave his name to the uncertainty principle.

439 p.140 — DENTALRINSE Austrian who identified blood groups, paving the way for safe transfusions.

440 p.62 — SEEDROT Dane who discovered the link between electricity and magnetism.

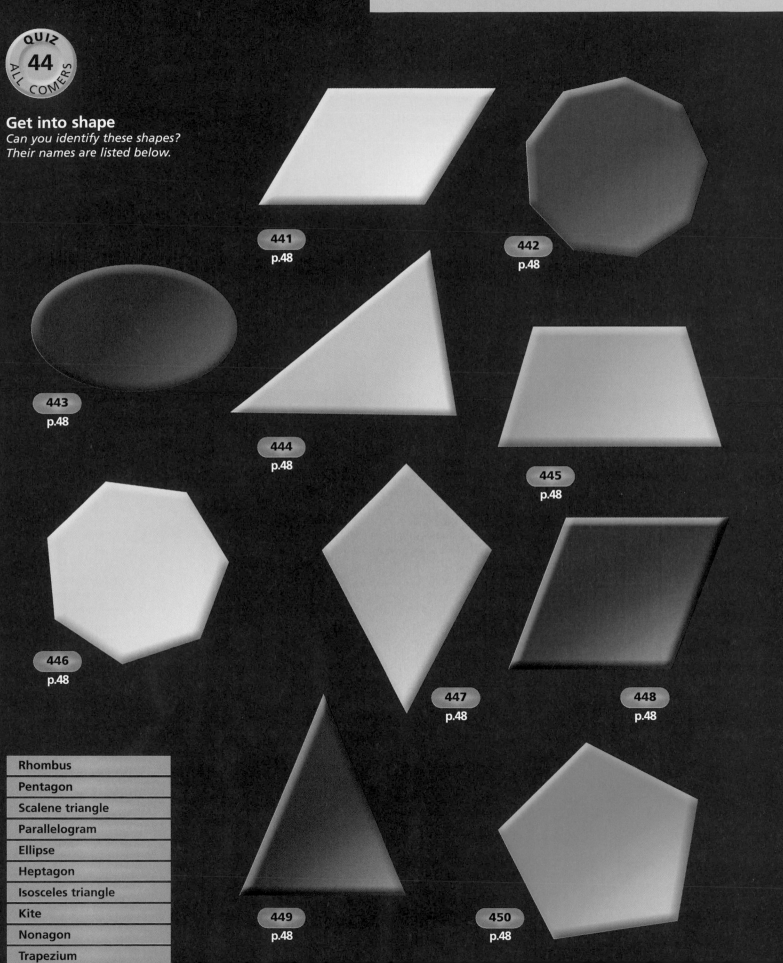

QUIZ
44
ALL COMERS

Get into shape
Can you identify these shapes?
Their names are listed below.

441 p.48

442 p.48

443 p.48

444 p.48

445 p.48

446 p.48

447 p.48

448 p.48

449 p.48

450 p.48

| Rhombus |
| Pentagon |
| Scalene triangle |
| Parallelogram |
| Ellipse |
| Heptagon |
| Isosceles triangle |
| Kite |
| Nonagon |
| Trapezium |

Mixed media
Multi-media puzzles.

451 p.50 In which 1995 mob movie did the producers gamble on Robert De Niro and Sharon Stone?

452 p.96 *Video Killed the Radio Star* was the first video played on which satellite television channel?

453 p.106 Which classic 1927 Fritz Lang movie features a robot called 'false Maria'?

454 p.96 In which children's book does an obnoxious boy called Mike Teevee get trapped inside a giant television set?

455 p.102 What are Helvetica, Geneva and Times New Roman?

456 p.144 Where did Georges Méliès take viewers in his pioneering 1902 science-fiction film *La Voyage dans la Lune*?

457 p.92 What does dot dot dot, dash dash dash, dot dot dot spell?

458 p.98 Which 1982 Disney movie used early computer animation to show a 'light bike' race in cyberspace?

459 p.142 The 1954 comedy classic *Doctor in the House* helped to launch the career of which actor, playing the part of medical student Simon Sparrow?

460 p.146 Which medium was explored by Arnold Schwarzenegger in the 1990 film *Total Recall*?

Testing your metals
Metal-based questions.

461 p.116 Which mineral, classified as a metal, is found in dairy products and is essential for healthy bones and teeth?

462 p.62 Do metals become superconductors at very high or very low temperatures?

463 p.78 Why is aluminium so popular with aeroplane manufacturers?

464 p.116 Which metal is vital to the blood's ability to carry oxygen?

465 p.116 Which metal is the only metal that is liquid at room temperature?

466 p.126 Which alloy, mainly tin and lead, was once used to make tankards and communion plates?

467 p.116 Which metal was named after the island of Cyprus?

468 p.116 Which metal is used to galvanise iron to protect it from rust?

469 p.126 Which three metals make up the alloy known as Alnico?

470 p.122 Which metal is chiefly obtained from an ore called bauxite?

Musical medley
All the answers are music related.

471 p.72 Which groundbreaking band lived 'in a yellow submarine'?

472 p.76 In 1978, which band was 'down in the Tube station at midnight'?

473 p.74 Which royal-sounding pop star sang 'Little Red Corvette' in 1983?

474 p.96 Who won critical acclaim in 1997 with the album *OK Computer*?

475 p.86 Which band's tenth album cover was black with a prism splitting light into the colours of the rainbow?

476 p.66 Which London power station graced the cover of the 1977 album *Animals*?

477 p.84 Which Frankie Goes to Hollywood video had Ronald Reagan in a fist fight with Soviet leader Konstantin Chernenko?

478 p.42 Which British band released the single 'Firestarter' in 1996?

479 p.104 In 1974, who sang the lyrics 'dynamite with a laser beam, guaranteed to blow your mind'?

480 p.72 Which pop star was 'Sailing' in 1975?

Pot luck
A mixed selection of teasers.

481 p.84 Which Renaissance artist is often credited with inventing the tank?

482 p.108 Who formed Lakeside Programming Group, a traffic management computer company, in 1968 at the age of 14?

483 p.72 Which Ancient Greek discovered the principle that explains why ships stay afloat?

484 p.110 What comprises both a dollar and a byte of data?

485 p.130 Which 'waste gas' breathed out by animals is used by plants to make food?

486 p.120 Which metal, with the chemical symbol Pb, is used to line containers for radioactive material?

487 p.150 In quantum physics, whose 'cat' is regarded as 50 per cent alive until it is observed?

488 p.72 What record is held by the VLCC *Jahre Viking*?

489 p.142 What do the red and white stripes on a barber's pole represent?

490 p.106 CAD is Computer Aided Design. What is CAM?

For answers and more facts go to the page given below each question number.
For quick answers to complete quizzes 45 to 51 go to page 156.

Plastic fantastic

Ten questions related to plastics.

491 p.128 Which early plastic was invented by the Belgian-born chemist Leo Baekeland?

492 p.128 What tax did Ireland impose on shoppers for environmental reasons in March 2002?

493 p.98 Which flammable plastic made from plant products revolutionised the film industry in the late 19th century?

494 p.122 Which natural resource provides the base material for most modern plastics?

495 p.142 Rhinoplasty is plastic surgery on which part of the body?

496 p.128 Which polyurethane fibre is used to make leggings and other close-fitting clothes?

497 p.128 What 'm' are organic molecules that combine to form polymers?

498 p.124 Neoprene is a synthetic form of which natural substance?

499 p.128 Which was the first country in the world to fully convert to plastic bank notes?

500 p.128 Which tough, acrylic thermoplastic is often used in place of glass?

Figure it out

A quiz on numbers and mathematics.

501 p.44 What is the smallest 'perfect' number?

502 p.44 After which Italian mathematician was the series beginning 0, 1, 1, 2, 3, 5, 8 … named?

503 p.44 What unreasonable-sounding name is given to numbers that cannot be expressed as exact fractions?

504 p.44 A googol is written as a 1 followed by how many zeroes?

505 p.44 Which cartoon family has a local cinema called the Googolplex?

506 p.48 If two sides of an isosceles triangle are 3 cm and 7 cm long, what length is the third side?

507 p.44 Which number was first used in the 7th century AD in India?

508 p.44 What is the name given to the ratio 1:1.618, which is seen in art and nature?

509 p.44 If one, two and three are cardinal numbers, by what term are first, second and third known?

510 p.46 What is one-half of two-thirds of three-quarters?

People and places

Select the correct option from the four possible answers.

511 p.122 Which country is the world's largest exporter of oil?

A Indonesia	B Iran
C Kenya	D Saudi Arabia

512 p.116 Who drew up chemistry's periodic table?

A Dmitri Mendeleyev	B Jean Lamarck
C Max Planck	D Carolus Linnaeus

513 p.92 In which country was the first long-distance telegraph laid?

A England	B Russia
C Germany	D The USA

514 p.144 Who invented the Japanese word *chindogu*, or 'weird tool'?

A Akira Kurosawa	B Kenji Kawakami
C Akio Morita	D Ryuichi Sakamoto

515 p.42 The oldest image of a wheel came from the Sumerian town of Uruk. In which modern country was Sumer?

A China	B Afghanistan
C Iraq	D Ethiopia

516 p.144 Which science-fiction writer wrote the books that inspired *Bladerunner* and *The Minority Report*?

A Michael Moorcock	B Arthur C. Clarke
C Iain M. Banks	D Philip K. Dick

517 p.76 Who built the first-ever steam locomotive?

A Robert Stephenson	B George Stephenson
C George Westinghouse	D Richard Trevithick

518 p.150 Whose law states that equal volumes of all gases at the same temperature contain the same number of molecules?

A Rutherford's Law	B Avogadro's Law
C Meitner's Law	D Thompson's Law

519 p.66 In which country was Robert Bunsen, inventor of the bunsen burner, born?

A Australia	B France
C Scotland	D Germany

520 p.128 Who invented polythene?

A Joseph Swan	B Reginald Gibson
C Thomas Edison	D Hubert Cecil Booth

QUIZ 52 WARM UP

Pot luck
A mixed selection of teasers.

521 p.42 Which was invented first, pottery or the wheel?

522 p.118 What is the common term for iron oxide?

523 p.106 Which films feature C3PO and R2D2?

524 p.134 Which is larger, a gene or a chromosome?

525 p.126 Which 'Age' began when people first mixed copper with tin?

526 p.40 Which science involves the study of elements and compounds?

527 p.128 Which is stronger, kevlar or steel?

528 p.130 Which gas makes up 78 per cent of the air at ground level?

529 p.50 If a fair coin is flipped twice and comes up tails twice, what are the chances that it will be heads next time?

530 p.132 Which have the more complicated structure, organs or tissues?

QUIZ 53 IN YOUR STRIDE

National identities
For most of these answers, simply name the country.

531 p.58 Which country did Hans Brinker supposedly save by plugging a leak in a dyke with his finger?

532 p.54 In which country was Albert Einstein born and in which did he die?

533 p.54 From which country did the physicist Mikhail Lomonosov hail?

534 p.44 Brahmaguptra was the first mathematician to incorporate zero into arithmetic. Which country was he from?

535 p.92 Which country was the first to popularise the fax machine?

536 p.128 Which novel by Scottish author Irvine Welsh has a title that sticks?

537 p.76 A golden spike driven into the track completed the first railway across which country?

538 p.148 The founder of quantum theory, Max Planck, hailed from which European nation?

539 p.148 In which capital city is the Nobel Museum?

540 p.134 What nationality was Gregor Mendel, who established the rules of genetic inheritance?

QUIZ 54 ALL COMERS

Encrypted energy
Use the clues to unravel the anagrams next to them.

541 p.70 ANEWREBEL Energy obtained from sustainable sources.

542 p.60 ITCYCLETIRE Phenomenon caused by the movement of electrons.

543 p.60 RUNALEC Energy released by the annihilation of matter.

544 p.60 DOSUN RARERIB Broken by objects travelling at more than 1230 km/h (765 mph).

545 p.60 LOOKJULIE A thousand standard units of energy or work.

546 p.60 COCONUTDIN Transfer of heat from one solid to another.

547 p.68 FENCAR The EU member with the highest proportion of electricity produced by nuclear power.

548 p.68 NEARGYM A European country which has pledged to phase out nuclear power entirely.

549 p.60 CUMINBOOST The conversion of chemical energy into heat and light.

550 p.60 LEAPTINTO The type of energy an object has because of its shape or position.

QUIZ 55 CHALLENGE

Digging deep
A series of underground questions.

551 p.58 The Simplon Orient Express uses the Simplon tunnel to pass between which two countries?

552 p.122 By what name is the glistering mineral iron pyrite better known?

553 p.58 Which civilisation dug the first road tunnels?

554 p.76 In which city is the world's oldest underground rail system?

555 p.122 Pitchblende is a source of which radioactive element?

556 p.58 The world's first known pedestrian tunnel ran under which river?

557 p.122 Why do people mine for quartz veins when quartz is plentiful in sand?

558 p.122 Which metal is extracted from ores called magnetite and haematite?

559 p.68 What is scheduled to be stored 500 m (1500 ft) underground at Yucca Mountain, Nevada?

560 p.122 Which diamond, dug from the Premier Mine in South Africa, weighed 3106 carats or 620 g (1 lb 6 oz)?

For answers and more facts go to the page given below each question number.
For quick answers to complete quizzes 52 to 56 go to page 156.

Through the looking glass

Can you work out what these pictures were taken with? Match the images to the instruments or technologies listed below.

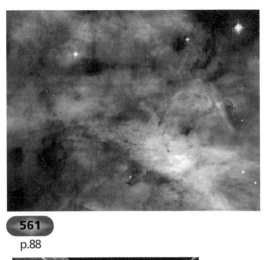

561
p.88

562
p.88

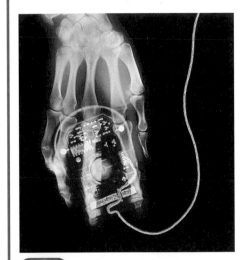

563
p.88

564
p.88

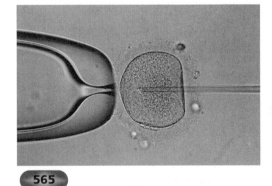

565
p.88

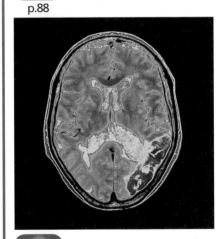

566
p.140

567
p.88

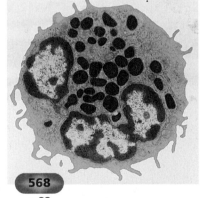

568
p.88

Atomic force microscope

Hubble Space Telescope

Light microscope

Magnetic resonance imager

Radio telescope

Refracting telescope

Scanning electron microscope

Thermograph

Transmission electron microscope

X-ray machine

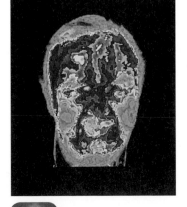

569
p.88

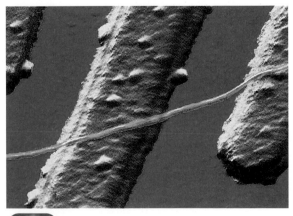

570
p.88

Household names

Identify the famous name or term described.

571 p.64 What name connects the vacuum cleaner to a dam, an FBI Director and a US President?

572 p.118 What substance containing ammonium carbonate is used to relieve faintness?

573 p.98 Which cartoon character first appeared in 1928 in a short feature titled *Steamboat Willie*?

574 p.118 By what name is the flavouring and food preservative diluted acetic acid better known?

575 p.108 Which computer operating system was launched by Microsoft in 1985?

576 p.64 Which inventor of cleaning equipment was inspired by watching a dust extractor in a factory roof?

577 p.110 Which home computer-games manufacturer has exclusive rights over the characters Mario and Donkey Kong?

578 p.112 What was the original, full name of the Internet Service Provider now branded simply as AOL?

579 p.124 Which US inventor, who later gave his name to a tyre company, came up with the process of vulcanisation?

580 p.128 By what three letters is polyvinyl chloride better known?

Fire and water

All questions are connected with either fire or water.

581 p.42 Who sang 'Light my Fire' in 1967?

582 p.72 What did the remotely operated vehicle *Argo* find at a depth of 3738 m (12 263 ft) in the Atlantic in 1985?

583 p.122 Which burns more readily, petrol or diesel?

584 p.152 Which temperature was defined in the 18th century using a mixture of salt water and ice?

585 p.66 'Carbon taxes' are imposed on countries supposedly to restrict the burning of what?

586 p.72 What is the name of the compartments on a submarine that are filled or emptied to alter depth?

587 p.114 What phenomenon of liquids allows insects such as pondskaters to stay afloat?

588 p.68 What nickname is given to deuterium oxide, a possible source of fuel for future nuclear reactors?

589 p.118 Which of the following three terms applies to the salt in salt water: solute, solvent or solution?

590 p.152 A kilogram of water is commonly called what?

Sporting chance

Answer the following sporting teasers.

591 p.124 In which athletics event do contestants use a 4-5 m (13-16 ft) long piece of fibreglass?

592 p.82 Epée and sabre are two types of sword used in Olympic fencing. What is the third?

593 p.82 Which light, throwing spear, now hurled on the athletics field, was standard issue for Roman legionaries?

594 p.126 Which Olympic medal has the name of an alloy?

595 p.90 Which cross-country sport requires the use of a map and a compass?

596 p.126 Pittsburgh's American football team was named for the production of which toughened metal?

597 p.124 Which sport is played with a puck made from vulcanised rubber?

598 p.60 When a sprinter bursts from the starting blocks, his or her potential energy changes to what other form of energy?

599 p.50 At what type of sporting event might you see a person signalling odds in tick-tack?

600 p.104 Which Olympic watersport has a class called Laser?

Double entendres

Which one word answers both parts of each question?

601 p.62 What 'r' is both a running race and a type of electronic switch?

602 p.110 What 'v' infects both people and computer programs?

603 p.84 Which guided missile was named after the brightest star in Ursa Minor?

604 p.132 What 'n' is the centre of an atom and the largest organelle in most animal and plant cells?

605 p.114 Which Malaysian car manufacturer shares its name with a type of subatomic particle?

606 p.118 Which play by Arthur Miller is also a cup used for heating chemicals?

607 p.136 Which liquor has the same name as a device for removing seeds from raw cotton?

608 p.48 What geometric term also describes a style of late medieval English Gothic architecture?

609 p.116 Which famous London theatre shares its name with a silvery-white metal?

610 p.152 Which ancient city had the same name as the weighing system for gems?

For answers and more facts go to the page given below each question number.
For quick answers to complete quizzes 57 to 63 go to page 156 and 157

Colour coded

Either the question or answer is linked to colour.

611 p.136 What was the name of the defoliant herbicide used by the US army in the Vietnam War?

612 p.118 What colour does litmus paper turn when dipped in acid?

613 p.118 And what colour does litmus paper become in an alkaline solution?

614 p.122 Which Texan fire-fighter led the operation to put out the Kuwait oil fires and dealt with the Piper Alpha disaster?

615 p.104 Which precious stone, normally red in colour, is used to produce the light in many lasers?

616 p.108 What did the IBM supercomputer Deep Blue become famous for playing in 1996?

617 p.86 Which two words link red, yellow and blue to John Travolta?

618 p.122 Which yellow element is extracted from the ground via the Frasch process?

619 p.122 What colour is the copper ore and semiprecious stone malachite?

620 p.78 What colour are 'Black Box' aeroplane flight recorders?

Pot luck

A selection of teasers on varied subjects.

621 p.72 Biremes had two and triremes had three decks of what?

622 p.82 Discounting bonus squares, which three-letter weapon would score 10 in Scrabble?

623 p.102 In which country was the world's oldest printed book, the *Diamond Sutra*, discovered?

624 p.110 Which computer games company was founded by Noel Bushnell, the inventor of Pong?

625 p.60 Which is the loudest: a gunshot, a rock concert or a space-rocket taking off?

626 p.144 Which prize was won in 1980 by the mathematical treatise *Gödel, Escher, Bach: an Eternal Golden Braid*?

627 p.140 What type of medical condition is indicated by the word-ending 'itis'?

628 p.150 Who predicted that microprocessors would double in complexity and power every two years?

629 p.152 How many bytes are there in a kilobyte?

630 p.114 Which fundamental particles have an electric charge of $+2/3$ or $-1/3$?

Gambler's luck

Select the correct option from the four possible answers.

631 p.50 Which of the following means money wagered on a bet?

A Pole	B Post
C Upright	D Stake

632 p.50 In what game can you win with a line or a house?

A Blackjack	B Poker
C Bingo	D Craps

633 p.50 What do the dots on opposite faces of a dice add up to?

A Six	B Seven
C Eight	D Nine

634 p.50 If you bet £10 on a horse that wins at odds of 100-1, how much money do you get back?

A £10	B £100
C £1000	D £1010

635 p.134 What decides whether a baby will be a boy or a girl?

A Father's sperm	B Mother's egg
C Date of conception	D Temperature of womb

636 p.50 If a country's death rate for a given year was 100, what percentage of the population died?

A One per cent	B Two per cent
C Five per cent	D 10 per cent

637 p.50 What are the chances of the ball landing on a particular number on a roulette table?

A One in 34	B One in 35
C One in 36	D One in 37

638 p.50 If you pick one set of numbers for all these national lotteries, which do you have the best chance of winning?

A Lotto (Ireland)	B Lotto (Great Britain)
C Translotto (Australia)	D The Big Game (USA)

639 p.50 What was Charles Fey's contribution to gambling?

A Founded Las Vegas	B Invented the slot machine
C Invented poker dice	D First online gambling site

640 p.50 Which of the following developed probability theory?

A August Möbius	B Blaise Pascal
C René Descartes	D Gottfried Leibnitz

QUIZ 64 WARM UP

QUIZ 65 IN YOUR STRIDE

QUIZ 66 IN YOUR STRIDE

QUIZ 67 ALL COMERS

Pot luck
A mixed selection of teasers.

641 p.46 What calculating device uses groups of beads threaded on rods?

642 p.82 In which European country was the bayonet invented?

643 p.146 NASA's 'Hyper-X' project aims to produce aircraft that will break which record?

644 p.114 Which state of matter has the greater density, liquid or gas?

645 p.90 Does sonar equipment use waves of high-frequency or low-frequency sound?

646 p.78 How did the Montgolfier brothers take to the air in 1783?

647 p.110 What sort of equipment do the letters PC stand for?

648 p.116 Which is bigger, an atom of hydrogen or an atom of tin?

649 p.148 In which century was the first Nobel prize awarded?

650 p.150 Which formula famously stated that matter and energy are equivalent?

Turn on, tune in
Ten questions on music and radio.

651 p.96 *Radio Gaga* was a comeback hit for which glam rock band?

652 p.96 In which century were the first radio broadcasts made?

653 p.76 Which train did Crosby, Stills and Nash sing about in 1969?

654 p.126 Which alloy of copper is used to make musical instruments?

655 p.72 What crossed the Mersey in Gerry and the Pacemakers' 1964 smash hit?

656 p.96 What is the main advantage to the listener of FM radio broadcasts over those broadcast on medium wave?

657 p.92 What type of person might know their 10-codes, have a handle and warn of bears?

658 p.64 What did Dire Straits have to install in their 1985 hit 'Money for Nothing'?

659 p.96 Which British company, named after a radio pioneer, saw its share value crash between 2000 and 2002?

660 p.96 What are KWAX, WFLY and WGMC?

Power to the people
Puzzles related to power sources.

661 p.122 Which Asian nation is the world's largest producer of coal?

662 p.66 Coal is a fossil fuel. What is it a fossil of?

663 p.122 Which TV family got rich when they discovered 'Black gold! Texan tea!'?

664 p.152 What is the standard unit of electric power?

665 p.122 Which cartel controls more than half of the global trade in oil?

666 p.122 What does pure natural gas smell of?

667 p.66 What fuel comes in forms called anthracite and lignite?

668 p.66 Which single country consumes 25 per cent of the world's electricity?

669 p.116 Which two metals are used in rechargeable 'NiCad' batteries?

670 p.64 Which uses the most electricity per minute: a hair-dryer, a three-bar electric fire, a computer or a television?

Measure for measure
All the answers are measurement related.

671 p.152 Which unit of measurement was based on the length of King Henry I's arm?

672 p.54 What changes when you are in a lift, your weight or your mass?

673 p.152 How many milligrams are there in a kilogram?

674 p.54 What is measured in metres per second squared?

675 p.152 Which social conflict is credited with bringing about the introduction of the metric system?

676 p.152 What are measured in radians, gradians and degrees?

677 p.152 How many pounds are there in an Imperial hundredweight?

678 p.152 How many feet are there in a mile?

679 p.114 In brewing, what is the name of the ratio that measures density relative to water?

680 p.152 What measure was originally derived from the distance from 'the Earth's Equator to the Pole divided by 10 million'?

For answers and more facts go to the page given below each question number.
For quick answers to complete quizzes 64 to 70 go to page 157.

Cutting edge recut
Use the clues to unravel the anagrams next to them.

681 p.146 — VIALRUT YETLIAR Technology that mimics the real world.

682 p.146 — JACKPET A portable device for travelling through the air.

683 p.146 — CRIBSOOT The science of machines capable of carrying out complex actions automatically.

684 p.136 — ICEGENT REIGNENGINE The alteration of organisms by DNA splicing.

685 p.74 — EDGYHORN An alternative to petrol, burned in fuel cells with water as a by-product.

686 p.146 — BLESSPODIA A throwaway term for single-use mobile phones.

687 p.146 — ALGEBRABODIED What all plastics may soon be, thanks to new technology.

688 p.146 — HOTELBOUT Nickname of Danish king Harald, adopted to describe a wireless communications technology.

689 p.146 — TESTCIGAR What the 'S' stands for in the abbreviation SDI.

690 p.80 — ITINERANTLOAN CAPES INTOAST A joint venture between several countries, to be completed in 2005.

Counter points
More mathematical teasers.

691 p.54 — What is calculated by multiplying gravity by mass?

692 p.44 — Which book of the Bible contains two population censuses of the Israelites?

693 p.50 — Which organisation can you join if your IQ is in the 98th quartile?

694 p.144 — What number appears in the title of Ray Bradbury's novel about firemen?

695 p.46 — What is 32 divided by a quarter?

696 p.110 — In computing, two nybbles make one what?

697 p.44 — Counting in ancient Babylon was done using the sexagesimal number system. What base does the sexagesimal system use?

698 p.102 — How many sheets of paper are there in a ream?

699 p.102 — And how many reams are there in a bundle?

700 p.102 — By what name is the 42-line Bible better known?

Gone but not forgotten
Select the correct option from the four possible answers.

701 p.136 — Who invented the horse-drawn seed drill?

A Cat Stevens	B Mike Oldfield
C Jethro Tull	D Fleetwood Mac

702 p.124 — What did glassmaker Edward Libbey create in 1893?

A Bulletproof glass	B Double glazing
C Lead crystal	D Fibreglass

703 p.126 — What did chemist Humphry Davy discover in 1807?

A Aluminium	B Helium
C Electricity	D Pluto

704 p.66 — Where was gas street lighting first used?

A New York	B Paris
C London	D Berlin

705 p.64 — What did Ambrose Fleming invent in 1904?

A The diode valve	B The electric oven
C Toothpaste	D Four-wheel drive

706 p.64 — Who invented the phonograph?

A Emile Berliner	B Thomas Edison
C John Logie Baird	D Alan Blumlein

707 p.124 — Where was the city of Tyre, famed in antiquity for its purple dye?

A Italy	B Turkey
C Greece	D Lebanon

708 p.126 — What is Invar, developed by Charles Guillaume in 1890?

A An iron-nickel alloy	B An early plastic
C A smokeless fuel	D A lubricating oil

709 p.124 — Who came up with the word rubber in 1770?

A Isaac Newton	B John Dalton
C Joseph Priestley	D Charles Lyell

710 p.64 — When did Alexander Bain patent his first fax machine?

A 1843	B 1899
C 1928	D 1954

Life in miniature
A quiz on small-scale inventions and organisms.

711 p.110 — Which miniature games system made by Nintendo has sold more than 100 million units?

712 p.108 — Which device, introduced in the 1970s, changed the way that maths is taught in schools?

713 p.140 — Which French scientist, forever associated with milk, demonstrated that microorganisms cause disease?

714 p.88 — What device, invented in the 17th century, enabled scientists to see microorganisms?

715 p.100 — Which tiny stereo, invented by Sony in 1979, brought a new way to listen to music?

716 p.74 — Which small car, first sold by the British Motor Corporation in 1959, was redesigned and relaunched in 2001?

717 p.108 — What tiny processing device is at the heart of every home computer?

718 p.88 — Which visual correction aids were conceived by Leonardo da Vinci but not widely available until the 1940s?

719 p.94 — Which now common pocket-sized devices were the size of a house brick when introduced in the 1980s?

720 p.132 — Which single-celled organisms with no cell nucleus are the smallest living things?

Surf's up
Internet-related questions.

721 p.112 — What do the letters www stand for in an Internet address?

722 p.112 — What are Alta Vista, Google and HotBot?

723 p.112 — Which British company runs the world's most popular information website outside the USA?

724 p.112 — What 'h' connects two separate files or web pages on the Internet?

725 p.112 — Which on-line retailing venture was founded by well-read entrepreneur Jeff Bezos?

726 p.112 — Which Hollywood actress had her identity stolen in the 1995 film *The Net*?

727 p.112 — What does the 'i' stand for in iMac?

728 p.112 — Which city-state's website is driven by servers called Raphael, Michael and Gabriel?

729 p.112 — What three letters at the end of an Internet address denote a web site for a not-for-profit concern?

730 p.112 — Why did the island of Tuvalu receive an unexpected cash windfall when it was allocated its Internet country code?

Going places
Ten questions on travel and transport.

731 p.76 — The world's longest railway line runs from Moscow to Vladivostok. By what name is it commonly known?

732 p.72 — What does ro-ro mean when used to refer to ferries?

733 p.80 — Who was the second man on the Moon?

734 p.76 — Japan's first shinkansen ('new trunk line') train ran between Tokyo and which other city?

735 p.78 — What are contrails, left by jet aeroplanes, made of?

736 p.72 — George Harbo and Frank Samuelsen were the first to row across which body of water?

737 p.76 — Where on a train would you find a bogie?

738 p.76 — Where would you find a funicular railway?

739 p.78 — Sir Frank Whittle and Dr Hans von Ohain are recognised as independent co-inventors of what?

740 p.78 — Why do the pilot and co-pilot on a passenger flight eat different types of food?

Pot luck
A mixed selection of teasers.

741 p.84 — General John T. Thompson gave his name to which wartime invention?

742 p.86 — Which of the seven colours in the visible light spectrum is omitted from many modern textbooks?

743 p.126 — Traditionally, which metal was always present in an amalgam?

744 p.114 — In which country is the CERN subatomic particle accelerator?

745 p.122 — What are Cora, Bray, Andrew, Tartan, Brent and Forties?

746 p.120 — In which shoot-em-up game for the PC and Playstation2 do you control scientist Gordon Freeman?

747 p.110 — The Turing Test is designed to test computers in which field of computer studies?

748 p.46 — Who, in the mid-1990s, completed his proof of Fermat's Last Theorem?

749 p.68 — What type of subatomic particles are absorbed by the control rods inside a nuclear reactor?

750 p.150 — How does the First Law of Thermodynamics differ from the Law of Conservation of Energy?

**For answers and more facts go to the page given below each question number.
For quick answers to complete quizzes 71 to 75 go to page 157.**

QUIZ **75** CHALLENGE

Inventions through the ages
When were these things invented?
Match the objects to the dates below.

| 2nd century BC |
| 1st century AD |
| c.1810 |
| 1876 |
| 1888 |
| 1901 |
| 1930 |
| 1938 |
| 1949 |
| 1971 |

Vacuum cleaner
751
p.64

Instant coffee
752
p.138

Digital watch
753
p.52

Canned food
755
p.138

Telephone
754
p.92

Clear adhesive tape
756
p.128

Bar code
757
p.104

Horseshoe
759
p.136

Record
758
p.100

Magnetic compass
760
p.62

QUIZ 76 WARM UP

Pot luck
A mixed selection of teasers.

761 p.104 In which science-fiction films do Jedi knights fight with light sabres?

762 p.114 Does water expand or contract when it becomes a solid?

763 p.90 What radio wave-based navigation system is used in air-traffic control?

764 p.78 What burst into flames at Lakenhurst, New Jersey, in 1937?

765 p.62 What does the 'c' stand for in DC electricity?

766 p.146 What, according to press reports in April 2002, was a patient of Dr Severino Antinori pregnant with?

767 p.94 What are the Siemens C45, Samsung T100 and Motorola V70?

768 p.144 Who wrote *The Restaurant at the End of the Universe* and *The Long Dark Tea-Time of the Soul*?

769 p.44 By what name is the number 3.14159 commonly known?

770 p.102 DTP software is used for desktop what?

QUIZ 77 IN YOUR STRIDE

First and last
All solutions begin and end with the same letter.

771 p.152 One thousandth of a gram.

772 p.72 Sea canoe made from sealskin stretched over a light wooden frame.

773 p.48 One of the three trigonometric ratios, along with sine and cosine.

774 p.68 Structure in which controlled nuclear reactions take place in order to release energy.

775 p.42 Continent where the oldest known pottery vessels were found.

776 p.130 Physical and chemical reactions that take place inside living organisms to produce energy and allow growth.

777 p.114 The largest type of subatomic particle.

778 p.44 A prime number and the square root of 361.

779 p.62 A conductor through which an electric current enters or leaves a battery.

780 p.56 A building term for a horizontal beam resting on two posts.

QUIZ 78 IN YOUR STRIDE

Messing with nature
Use the clues to unravel the anagrams next to them.

781 p.130 ROUGESHEEN SAGES Substances responsible for global warming.

782 p.130 EPSOMCODE What dead organic matter does, with the help of bacteria and fungi.

783 p.132 ISETUS Body structure made entirely from cells of one type.

784 p.130 RINGTONE A gas present in Earth's atmosphere that must be chemically converted before plants can use it.

785 p.130 ADOPTANITA A change of behaviour or form to reflect changes in the environment.

786 p.130 THOSESHINYTOPS The reason that there is enough oxygen on the planet for us to breathe.

787 p.134 SMOOCHMORE A giant molecule of DNA coiled around a protein core.

788 p.130 INDUCETORPOR One of the six characteristics common to all life.

789 p.132 TRASHYBROCADE A group of substances that includes starches and sugars.

790 p.132 COSTAMPLY A jelly-like material that makes up the bulk of animal and plant cells.

QUIZ 79 ALL COMERS

Compound interest
Ten questions on elements, compounds and reactions.

791 p.118 Which two chemical elements go to make up water?

792 p.132 Which three elements are found in all carbohydrates?

793 p.118 What is always released when chemical bonds are broken?

794 p.134 Which vital substance contains adenine, cytosine, guanine and thymine?

795 p.118 What 'c' is a substance that speeds up a chemical reaction?

796 p.118 Which is stronger, an ionic or covalent bond?

797 p.132 Which complex organic compounds are made up of amino acids?

798 p.122 Which carbon-based compound is the main component of natural gas?

799 p.118 What two new substances are formed when sodium is added to water?

800 p.118 The suffix '-ate' on the name of a chemical compound signifies the presence of what?

Ps and Qs
Every answer begins with the letter p or q.

801 p.142 What 'p' is a device used to stimulate heart muscle and regulate the pulse?

802 p.52 What 'q' is a form of silica used to keep time in watches and clocks?

803 p.152 What 'p' links pressure to computers and a French mathematician?

804 p.82 What 'q' is a vessel for carrying arrows?

805 p.94 What 'p' is a portable radio device that bleeps to alert its wearer of incoming messages?

806 p.40 What 'p' is the science of properties and interactions of matter and energy?

807 p.138 What 'q' is a meat substitute made from edible fungus?

808 p.116 In chemistry, which element has the symbol P?

809 p.86 What 'p' is the word for a particle of light?

810 p.86 Particles of energy are collectively known by what 'q'?

The creative process
A quiz on growth, production and invention.

811 p.130 What type of organisms need light in order to make food?

812 p.52 What seem to be melting in Salvador Dali's painting *The Persistence of Memory*?

813 p.106 Who made cars more affordable by installing a moving assembly line?

814 p.68 Which process that powers the Sun offers the hope of clean energy on Earth?

815 p.42 Apart from transport, what was the wheel first used for?

816 p.126 Which other metal is added to give strength to sterling silver?

817 p.106 What thread-making device did James Hargreaves name after his daughter?

818 p.126 Which mystical object did alchemists believe would turn base metals into gold?

819 p.128 What did Jacques Brandenburger create from cellulose in 1908?

820 p.42 How old is the oldest known textile, to the nearest 500 years?

Right tool for the job
Select the correct option from the four possible answers.

821 p.76 What are the objects that link rails together known as?

A Snoozers	B Nappers
C Dozers	D Sleepers

822 p.70 Photoelectric cells generate electricity using what?

A Sunlight	B Wind
C Water	D Nuclear fuel

823 p.90 What does GPS stand for?

A Greek Postal Service	B Grid Point Standard
C Ground Pinpoint Satellite	D Global Positioning System

824 p.42 Eoliths are ancient tools made of what?

A Wood	B Stone
C Iron	D Bronze

825 p.120 What is measured using a Geiger counter?

A Temperature	B Radioactivity
C Acidity	D Speed

826 p.78 Why do stunt pilots prefer biplanes to monoplanes?

A Faster	B More manoeuvrable
C More economical	D Safer

827 p.46 Napier's bones were a set of ivory sticks used to simplify what type of calculation?

A Addition	B Subtraction
C Multiplication	D Division

828 p.90 Germany's *Knickebein* system directed bombers through darkness in the Second World War. What did it use?

A Spotlights	B Ultrasound
C Radio beams	D Microwaves

829 p.90 What type of person would use an ephemeris?

A An astronomer	B A weather forecaster
C An architect	D An engineer

830 p.90 Which of the following once used sextants?

A Chemists	B Farmers
C Doctors	D Sailors

True or false?
Decide whether the given facts are correct or not.

831 p.90 A gyrocompass always points to magnetic north.

832 p.144 H.G. Wells wrote *20 000 Leagues under the Sea*.

833 p.120 There is radiation all around us.

834 p.114 Most of the matter in the Universe cannot be seen or detected with light.

835 p.150 One watt of power is delivered by one amp of current flowing over a component with the potential of one volt.

836 p.120 Most radioactive elements have low atomic weights.

837 p.66 Norway, Paraguay and Zambia generate over 99 per cent of their electricity from renewable sources.

838 p.92 VHF stands for Very High Frequency.

839 p.146 In Japan you can buy robot fish.

840 p.104 Laser beams are always blue.

Feeling the heat
A hot and cold quiz.

841 p.60 Which boils at a higher temperature, salt water or fresh water?

842 p.42 What type of stone was used both for early weapons and as a means of creating fire?

843 p.42 Smelting is the use of heat to extract what from an ore?

844 p.86 What kitchen appliance heats food by causing the molecules inside it to vibrate?

845 p.66 Which capital city is heated entirely by geothermal energy?

846 p.60 Which common liquid is used in 'minimum temperature' thermometers because of its low freezing point?

847 p.116 Which are better heat conductors, metallic or non-metallic elements?

848 p.152 Which temperature scale, named after a British baron, begins at absolute zero?

849 p.60 What is unusual about the temperature 40 degrees below zero?

850 p.126 What 'a' means to toughen metal by a process of gradually heating and cooling?

It's a date
Famous days and occasions.

851 p.44 What do the letters M, D, C, L, X, V and I have to do with the Great Fire of London?

852 p.52 What date occurred in 2000 and will occur again in 2400 but not in 2100, 2200 or 2300?

853 p.148 Why are the Nobel prize ceremonies held on December 10?

854 p.68 In July 1955, Arco, Idaho, became the first US town to have all of its electricity generated by what method?

855 p.52 On which day of the year does 'Twelfth Night' fall?

856 p.72 What was special about the return of the *Victoria* to Spain in 1522?

857 p.78 What did Louis Blériot fly across in 1909?

858 p.52 The period AD 2003-4 is equivalent to AH 1424 in which calendar?

859 p.64 In which decade was the domestic refrigerator introduced?

860 p.50 What is El Gordo ('The Fat One'), an event run every year in Spain on December 22?

Pot luck
A mixed selection of teasers.

861 p.76 What are the Métro, MTR and Eidan?

862 p.96 What do the letters FM stand for when applied to radio broadcasts?

863 p.114 Which New Zealand scientist first split the atom in 1919?

864 p.114 What 'p' is the fourth state of matter?

865 p.78 What set the Heinkel He178 apart in 1939?

866 p.90 What was known for more than two decades as Asdic before being renamed by the US navy in the Second World War?

867 p.128 In which European country is the plastic explosive Semtex made?

868 p.70 What seven-letter word means organic matter or waste as a source of renewable energy?

869 p.66 To the nearest 5 per cent, how much of the world's electricity is generated by burning fossil fuels?

870 p.120 Which Nobel laureate mother and daughter both died of leukaemia after years of working with radioactive materials?

For answers and more facts go to the page given below each question number.
For quick answers to complete quizzes 83 to 87 go to page 158.

QUIZ
87
IN YOUR STRIDE

Luminary line-up
How many of these famous scientists could you pick out from a crowd? Match the names to the faces of their owners.

Archimedes

Tim Berners-Lee

Rachel Carson

Marie Curie

Albert Einstein

Galileo Galilei

Isaac Newton

Edwin Hubble

Guglielmo Marconi

Craig Venter

871 p.72

872 p.96

873 p.54

874 p.134

875 p.112

876 p.148

877 p.88

878 p.54

879 p.88

880 p.136

Study time
Match the descriptions to the names of the sciences.

881 p.40 The science of number, quantity and shape.

882 p.40 The scientific study of plants.

883 p.40 A branch of the life sciences concerned with organisms too small to see with the naked eye.

884 p.40 The field of study concerned with long molecules, particularly those that form plastics.

885 p.40 The study of inherited characteristics.

886 p.40 The science of chemical processes that occur inside living organisms.

887 p.40 The study of the interaction between organisms and their surroundings.

888 p.40 A discipline that seeks to understand long-term weather patterns.

889 p.40 A field of geology concerned with earth movements.

890 p.40 A branch of physics concerned with heat and its relation to other forms of energy.

Biochemistry
Botany
Climatology
Ecology
Genetics
Mathematics
Microbiology
Polymer chemistry
Seismology
Thermodynamics

Odd one out
Pick the odd one out in each of the following lists.

891 p.108 Mouse, keyboard, record deck, monitor.

892 p.120 Alpha, beta, gamma, epsilon.

893 p.40 Geometry, trigonometry, probability, microbiology.

894 p.66 Wood, oil, natural gas, coal.

895 p.114 Proton, carbohydrate, neutron, electron.

896 p.136 John Deere, Massey Ferguson, David Brown, Harley-Davidson.

897 p.128 Cotton, terylene, orlon, rayon.

898 p.100 Tweeter, barker, woofer, subwoofer.

899 p.94 Mobile phone, fax machine, television, pager.

900 p.100 A-type, B-type, E-type, S-type.

Nouvelle cuisine
Mainly food-related teasers.

901 p.126 What would you do with EPNS – clean your teeth with it, eat with it or drive it?

902 p.138 What type of substance is tartrazine, linked to hyperactivity in children?

903 p.128 On what kitchen utensils might you find polytetrafluoroethylene?

904 p.138 Who in the 1960s famously created a series of images based on a tin of Campbell's condensed soup?

905 p.138 Why have CFCs been replaced by other chemicals in fridges and freezers?

906 p.138 What metallic element is the main constituent of most 'tin' cans?

907 p.138 MSG is a flavouring best-known for its use in Chinese cooking. What does MSG stand for?

908 p.138 What form of food poisoning is caused by a toxin that is also used to reduce wrinkles?

909 p.138 What did Frederick Tudor export from New England in the 1800s, enabling people to keep food fresh for longer?

910 p.138 What is the English name for the processed fish product known to the Japanese as *surimi*?

Digital resolution
Questions with a digital connection.

911 p.108 Which digit is above the six on a standard calculator keypad?

912 p.92 Which digit is above the six on a telephone keypad?

913 p.142 Which flowering plant yields digitalis, a drug used to stimulate heart muscle?

914 p.100 What do the letters DVD stand for?

915 p.100 What was the first fully digital music reproduction medium available to the public?

916 p.112 What sort of file might be sent by email as a JPEG attachment?

917 p.96 The images on a digital television are produced by numerous 'picture elements', which are better known as what?

918 p.98 What was the first feature film made entirely by digital animation to be shown in cinemas worldwide?

919 p.98 On a traditional camera the image is displayed through a viewfinder. How is it displayed on a digital camera?

920 p.108 Used in computers, the American Standard Code for Information Interchange (ASCII) is also known by which two words?

For answers and more facts go to the page given below each question number.
For quick answers to complete quizzes 88 to 95 go to page 158.

Testing times
A time-related quiz.

921 p.52 Which movie starring Michael J. Fox featured a car with the licence plate 'OUTATIME'?

922 p.52 Which day of the week took its name from the Norse god of thunder?

923 p.52 What agricultural-sounding name is given to the full moon nearest the autumn equinox?

924 p.120 How is the radioactive substance carbon-14 useful to archaeologists?

925 p.52 What type of clock, also called a clepsydra, was used to measure the lengths of speeches in ancient Athens?

926 p.88 On which planet in the Solar System would the phrase 'it's been a long day' be most appropriate?

927 p.52 What timekeeping device was Galileo inspired to invent after watching a priest swinging an incense burner in church?

928 p.52 Sidereal time is time measured in relation to the apparent motion of what?

929 p.52 Concorde leaves London at 3pm and takes three hours to fly to New York. At what local time will it arrive?

930 p.52 The Long Now Foundation, which exists to promote long-term thinking, is building a clock that will run for how long?

Crossed wires
Use the clues to unravel the anagrams next to them.

931 p.112 ENTERTIN The global network of telephone lines and computers.

932 p.92 GREATELM A message decoded from telegraph and delivered by hand.

933 p.92 SANDCLOT The country where Alexander Graham Bell was born.

934 p.92 EASELSTILT Devices that bounce signals from one part of the world to another.

935 p.92 ERNESTVICAR A combined transmitter and receiver.

936 p.94 RULECALL A type of network that services mobile phone users.

937 p.92 BRIEF-TOPIC A cable that transmits light rather than electricity.

938 p.94 DIRTH NEATREGION The term for mobile phones that incorporate WAP technology.

939 p.94 TIEDRAGNET What the 'I' stands for in ISDN.

940 p.92 SEETHATNOW The name of the Briton who co-invented the telegraph system.

Pot luck
A mixed selection of teasers.

941 p.60 What 'c' is the process by which heat travels through water?

942 p.110 Which former playing-card manufacturer broke into the computer-game market with the Famicom console?

943 p.84 Why are the UK, USA, Russia, France and China the five permanent members of the UN Security Council?

944 p.108 What award did *Time* magazine bestow upon the computer in 1982?

945 p.82 What primitive weapon is mentioned in Samuel 1, Chapter 17 of the Bible?

946 p.118 What type of acid is vitriol, carried by Pinkie in Graham Greene's *Brighton Rock*?

947 p.66 What name was given to miners from north-east England, after they chose lamps designed by George Stephenson?

948 p.110 Computer programs are software and computer equipment is hardware. What is wetware?

949 p.148 Who presents the medal and diploma for five of the six Nobel prizes?

950 p.102 A sans serif typeface does not have any serifs on its letters. What are serifs?

Current affairs
Ten electricity-based questions.

951 p.62 On a circuit diagram, what illuminating device is marked by a cross in a circle?

952 p.62 Which French physicist gave his name to the SI base unit of electric current?

953 p.64 Which invention by Alexander Graham Bell made treasure hunting a popular pastime?

954 p.64 What 'c' ensures that a dynamo delivers direct current (DC) electricity?

955 p.108 What single word describes a resistor that transmits different voltages?

956 p.62 Which devices for storing electrical charge used to be called condensers?

957 p.62 What is the effective resistance of two 100 ohm resistors in a parallel circuit?

958 p.62 When fitting a plug, what would the formula 'power divided by voltage' help you to decide?

959 p.62 Which subatomic particles carry electricity through solids?

960 p.62 What 's' is a cylindrical coil of wire that becomes a magnet when electrical current is passed through it?

On at the end
All of the following describe things that end with 'on'.

961 p.46 Taking one number away from another.

962 p.42 A metallic element that lent its name to an age.

963 p.136 A plant grown solely to produce fibres for textiles.

964 p.108 A semiconductor that has a Californian valley named after it.

965 p.132 The chemical element that forms the structural basis of every lifeform on Earth.

966 p.120 A radioactive gas often associated with granite.

967 p.42 The process of directing water onto crops.

968 p.46 The physicist who developed calculus, better known for his work on gravity.

969 p.50 Any group of people or items studied by statisticians.

970 p.130 Breaking down sugars in the body to release energy; also another word for breathing.

Pot luck
A mixed selection of teasers.

971 p.110 What was the name of the computer that achieved consciousness in Arthur C. Clarke's *2001: A Space Odyssey*?

972 p.84 What was developed in the Manhattan Project?

973 p.82 Which was invented first, the sword or the shield?

974 p.150 Whose school report was said to have contained the comment 'You will never amount to very much'?

975 p.126 What man-made metal objects do numismatists collect and study?

976 p.132 In what part of a plant or animal cell is DNA stored?

977 p.70 The world's largest tidal power plant is on the Atlantic coast of which European country?

978 p.90 For what military purpose was sonar developed?

979 p.90 In which decade was the first satellite navigation system created?

980 p.114 Which subatomic particle did James Chadwick discover in 1932?

Mostly medicine
Ten medical questions.

981 p.140 Coryza and rhinitis are technical names for which inflammation of the nasal mucous membrane?

982 p.118 Which halogen compounds are added to some water supplies to prevent tooth decay?

983 p.142 Which anaesthetic was administered to Queen Victoria during the birth of her eighth child?

984 p.142 Originally obtained from willow bark, acetylsalicylic acid is better known by which common name?

985 p.142 Why did Louise Brown become famous on July 25, 1978?

986 p.140 In 1953, US doctor Jonas Salk created the first vaccine for which crippling disease?

987 p.148 Who, in 1903, became the first-ever woman to win a Nobel prize?

988 p.142 Which Swiss psychiatrist introduced the so-called 'ink-blot test'?

989 p.140 What 'g' describes an enlarged thyroid gland resulting from an iodine deficiency?

990 p.140 An ECG is a type of heart monitor. What does ECG stand for?

The final frontier
A quiz relating to space travel.

991 p.80 What crashed into the Pacific Ocean on March 23, 2001?

992 p.144 Jean Luc Picard, Benjamin Sisko, Kathryn Janeway, Jonathan Archer. Who's missing?

993 p.80 Which space probe has taken Bach's *Well-Tempered Clavier* out of the Solar System?

994 p.80 The Climate Orbiter and Polar Lander were lost on failed missions to which planet?

995 p.80 How many space shuttles were built for use outside the Earth's atmosphere?

996 p.80 What are GOES-WEST, GOES-EAST, GMS and Meteosat?

997 p.80 Which planet was the target of the NASA missions *Magellan* and *Mariner 5*?

998 p.80 What happened to Pluto from 1979 to 1999 and will happen again from 2207 to 2227?

999 p.80 The space probe *Galileo* observed active volcanoes on which of Jupiter's moons?

1000 p.80 NASA developed pens which write in zero gravity. What did Russian cosmonauts use instead?

For answers and more facts go to the page given below each question number.
For quick answers to complete quizzes 96 to 99 go to page 158.

THE WORLD OF
SCIENCE
AND
TECHNOLOGY

Essential facts, figures and
other information on The World of Science
and Technology

Branches of science

Core facts ❶

◆ Science is systematic, formulated knowledge. **Pure science** refers to knowledge of subjects that can be verified by proof or study.
◆ Until about the 18th century in Europe, pure science was considered a branch of philosophy, or even theology.

◆ **Technology** refers to the practical applications of scientific knowledge, and can be considered as old as the first human tool.
◆ Modern technology covers areas such as manufacturing, computers, electronics, communications, transport and engineering.

Mathematics ❷

Mathematics is the science of number, quantity and shape. It is divided into **pure mathematics** (mathematical theory for its own sake) and **applied mathematics**.

▶ **Algebra** is the study of the general properties of numbers, using letters and symbols for unknown or variable numbers.
▶ **Arithmetic** is the study of numbers and their relationships.
▶ **Calculus** deals with the study of continuously changing variables.
▶ **Geometry** is the study of space or area, using two-dimensional and three-dimensional shapes and their properties.
▶ **Probability** expresses the likelihood of a given event occurring.
▶ **Set theory** is the study of defined groups of entities and the use of mathematics to describe their properties and relationships.
▶ **Statistics** is the science of the collection, organisation and interpretation of numerical data.
▶ **Trigonometry** is the mathematical study of triangles and their properties.

Counting and simple calculation are natural to all humans, and the scientific study of mathematics dates back to the ancient Middle East, Egypt and Greece. Modern developments have included probability and statistics, with direct applications in commerce. Mathematics provides a precise language for expressing many scientific and technological concepts.

ELECTRICAL CIRCUIT The circuit board of a computer includes diodes, resistors and silicon chips.

Physics ❸

Physics is the science of the properties and interactions of matter, motion and energy. Its **laws** are usually expressed in mathematics and underlie all the other natural sciences.

▶ **Acoustics** is the science of sound.
▶ **Atomic physics** is the study of atoms, their parts and behaviour.
▶ **Electromagnetics** is concerned with the closely related forces of electricity and magnetism.
▶ **Mechanics** (or **classical mechanics**) is the study of the effects of forces on objects, other than subatomic particles or objects moving at near the speed of light or subject to intense gravity.
▶ **Optics** is the study of light and its properties, including its wave nature, reflection and refraction as a ray.
▶ **Particle physics** is the study of the fundamental subatomic constituents of matter.
▶ **Quantum mechanics** covers the behaviour of subatomic particles, objects moving at near the speed of light and objects subject to intense gravity.
▶ **Solid-state physics** looks at the properties of solid materials, including their conductivity.
▶ **Thermodynamics** is the study of heat and its relation to other forms of energy.

For most of history, physics was known as **natural philosophy,** and was concerned with understanding the observable world. In the 20th century, new discoveries within the atom and in space led to **quantum physics** and ongoing efforts to combine the two systems.

Chemistry ❹

This is the science of the composition of substances and the changes they undergo through interaction (chemical **reaction**). It involves the study of **elements** (made up of a single type of atom) and **compounds** (made up of two or more different types of atoms).

▶ **Biochemistry** is the study of the chemical processes that take place within living organisms.

▶ **Inorganic chemistry** is the study of the chemistry of all elements and compounds other than those containing carbon.

▶ **Organic chemistry** deals with the chemistry of compounds containing carbon, which are the basis of all living things.

▶ **Physical chemistry** is the study of the physics of chemical reactions, including the effects of heat or pressure and the movement of molecules.

▶ **Polymer chemistry** deals with compounds made up of large molecules in which a group of atoms is repeated. Some polymers occur naturally. There are many artificial polymers, also known as plastics.

Chemistry developed from the medieval study of **alchemy**, which sought to understand and change natural substances. Today, chemistry has practical applications in a huge range of products from plastics to medicines to fabrics to food, and the chemical industry is one of the largest in the world.

Life sciences ❺

These deal with living things and their life processes. Branches include: **bacteriology** (the study of bacteria); **botany** (plants); **entomology** (insects); **ichthyology** (fish); **ornithology** (birds); and **zoology** (animals). Other life sciences are shown below.

▶ **Ecology** is the study of the relationship between living things and their environments.

▶ **Embryology** is the study of human and animal lifeforms before their birth.

▶ **Genetics** is the science of heredity and variation in living things.

▶ **Microbiology** is the study of the biology of microscopic organisms.

▶ **Morphology** is the study of the form and structure of animals and plants.

▶ **Physical anthropology** is the study of evolution and variation in humans.

▶ **Physiology** is the study of the activities and functions of living organisms or their parts.

▶ **Taxonomy** is the science of the classification of plants and animals according to their features.

The science of biology dates from the accurate observations of the ancient Greeks in the 5th century BC, and arguably even earlier. The life sciences today are increasingly concerned with genetics (as in the **Human Genome Project**) and microbiology, and with ecology. Agriculture and medicine are major practical applications of the life sciences.

Earth sciences ❻

The earth sciences are those concerned with the study of the origin, structure and physical phenomena of the Earth (known as **geological** sciences), its waters (**hydrological** sciences) and also its atmosphere (**atmospheric** sciences).

▶ **Aeronomy** is the study of the Earth's upper atmosphere, including such phenomena as the auroras and magnetospheric storms.

▶ **Climatology** is the science of long-term changes in the Earth's weather systems.

▶ **Geochemistry** is the science of the Earth's chemical composition and changes that occur in this.

▶ **Geophysics** is the study of the physical properties of the Earth (such as gravity) and their changes.

▶ **Glaciology** is the study of glaciers and the Earth's ice caps.

▶ **Hydrology** is the study of water that lies close to the surface of the land, such as rivers.

▶ **Limnology** is the study of the Earth's lakes and inland seas.

▶ **Meteorology** is the study and prediction of short-term weather.

▶ **Oceanography** is the study and mapping of seas and oceans.

▶ **Physical geology** includes: **geomorphology** (the study of landforms); **mineralogy** (minerals); **palaeontology** (fossils); **petrography** and **petrology** (rocks); **stratigraphy** (layers of sediment); and **structural geology** (Earth structures).

▶ **Seismology** is the study of earthquakes.

Knowledge of the Earth's natural processes was used from earliest times. As sciences, however, geology and the related disciplines only emerged in the 18th century. Today, they underlie mining and mineral extraction, weather forecasting, and many aspects of ecology.

Early technology

Core facts ❶

◆ Cave paintings from around 20000 years ago show hunters using **spears** and **bows**.

◆ **Stone tools** and **fire** enabled the earliest humans to survive in the harsh environment of the Ice Ages and Interglacial periods.

◆ **Agriculture**, **pottery** and **textiles** are closely connected with the emergence of the first human civilisations after the last Ice Age.

◆ **Metalworking** and the development of the **wheel** made possible the technological advances of ancient Egypt, Greece and Rome.

◆ The **Stone**, **Bronze** and **Iron Ages** are named by archaeologists for the characteristic materials used for tools and weapons in these periods.

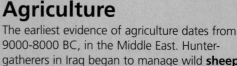

Agriculture ❷

The earliest evidence of agriculture dates from 9000-8000 BC, in the Middle East. Hunter-gatherers in Iraq began to manage wild **sheep** at this time. **Rice** cultivation began in China in about 6500 BC, and agriculture spread across southern Europe, Egypt and India from about 6000 BC. **Maize** was being cultivated in the Americas by about 3500 BC.

EGYPTIAN FARMER A harvesting scene on papyrus dating from around 1250 BC.

Pottery ❸

The **hardening of clay** by deliberately placing it near a fire can be traced back to early humans around 30000 years ago. The first known pottery vessels are from the **Jomon** culture of Japan, from about 12000 years ago. Pottery-making emerged in North Africa about 9500 years ago, in Asia Minor (Turkey) and southern Europe about 9000 years ago, in India about 8000 years ago, and in the Amazon Basin about 6000 years ago.

IRON-AGE POTTERY This clay feeding vessel in the shape of an animal was made in Austria in the 7th century BC.

The wheel ❹

The **earliest surviving image** of a wheel is in a Sumerian pictograph (writing symbol) from Uruk (Mesopotamia), dated about 3500 BC, which shows a wheeled sledge. Wheels found in graves in the Middle East date from a similar time. The first wheels were cut in a single piece from a log. **Spoked** wheels were developed in about 2000 BC in Asia Minor, leading to the invention of the chariot. Wheels were also used for making **pottery** from about 3500 BC in Mesopotamia. The first mention of a **water wheel** to provide power for grinding corn dates from about 85 BC, in Greece.

PAKISTAN ARTEFACT A terracotta model from around 2600 BC of a two-wheeled cart strongly resembles farm vehicles still in use to this day in the area.

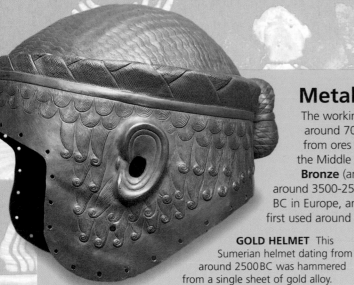

Metalworking ❺

The working of pure ores began with copper around 7000 BC in the Balkans. **Smelting** metals from ores and **casting** were both developed in the Middle East about 6200 BC.

Bronze (an alloy of copper and tin) was first used around 3500-2500 BC in the Middle East, 2500-2000 BC in Europe, and around 2000 BC in China. **Iron** was first used around 1500-1000 BC in Asia Minor, in the first millennium BC in Europe, and around 600 BC in China. **Copper** and **gold** were used in Peru and parts of North America from the 1st century AD.

GOLD HELMET This Sumerian helmet dating from around 2500 BC was hammered from a single sheet of gold alloy.

342

Fire

Hominids (upright-walking human ancestors) in Africa had learned the controlled use of fire over 1.4 million years ago. However, a reliable means of making fire (from **flint**) was not found until about 9000 years ago.

Fire was used for warmth, cooking and driving animals during hunting. It became important for firing kilns and furnaces and, as agriculture developed, for clearing underbrush.

Spinning and weaving ❻

The spinning of animal or vegetable fibres into **thread** dates from about 7000 BC, from the Middle East; the earliest tools used were a **distaff** (a cleft stick holding a bundle of fibres) with a weighted **spindle** to twist them together. The oldest known **textile**, from Asia Minor, dates from around 7000 BC. The first **looms** for weaving wool date from about 5000 BC.

FAMILY SCENE A Hittite lady spins wool in a bas-relief from the 8th century BC.

Tools and weapons ❼

The oldest known **stone** tools were made by *Homo africanus* in Ethiopia about 2.4 million years ago. The first **tool-making culture** was that of *Homo habilis*, in East Africa's Rift Valley. The oldest known wooden tool (a **spear**) was found in Shöningen, Germany, dating from about 400 000 years ago.

The earliest tools were flaked stone **blades** and **choppers** for butchering meat. Over 100 distinct stone and bone tool types are known from the period of modern human colonisation of Europe (around 40 000 years ago), including **needles** and **arrowheads**.

STONE-AGE TOOLS Arrowheads were made by striking off fragments of flint from flintstone.

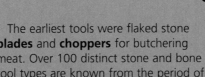

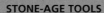

TIMESCALE ❽

▶ **8000 BC** Wheat, barley, sheep and goats are all domesticated in the Middle East; pigs are domesticated in south-east Asia.

▶ **7000 BC** Copper is first used in tools in Asia Minor and the Balkans.

▶ **6000 BC** Irrigation begins in Mesopotamia; pottery and domesticated plants (including rice) and animals appear in China.

▶ **4500 BC** The plough is first used in Europe.

▶ **4000 BC** Copper is first used in Egypt and India; pottery is being made in the Amazon Basin.

▶ **3500 BC** The potter's wheel is in use in India; the cartwheel is in use in India and Mesopotamia. In the Americas, maize is widely cultivated.

▶ **3000 BC** Bronze is first produced in Egypt and Mesopotamia; the potter's wheel is introduced to China.

▶ **2640 BC** The Chinese silk industry begins.

▶ **2300 BC** The first pottery is produced in Central America; bronze is in use in the eastern Mediterranean.

▶ **2250 BC** The first known irrigation dam is built.

▶ **2000 BC** Bronze comes into use in India and China.

▶ **1440 BC** The first known metalwork is carried out in South America.

▶ **1400 BC** Iron ploughshares are first used in India.

▶ **1350 BC** War chariots are introduced to China.

▶ **1200 BC** Agriculture spreads through North America.

▶ **1000 BC** Iron comes into use around the Mediterranean .

▶ **600 BC** Iron is used for the first time in China.

SUMERIAN WHEELS Solid-wheeled carts appear in this detail of a Sumerian battle scene of around 2600 BC.

Number systems

Core facts ❶

◆ The study of numbers and their patterns is today known as **number theory**.
◆ The earliest numbers were **natural** numbers and fractions, but the ancient Greeks also recognised certain **irrational** numbers (those which cannot be expressed as exact fractions).
◆ It is believed that **zero** was first used as a number in India in about AD 628.

Number systems ❷

The Hindu/Arabic system universally used today was slowly adopted in Europe from the 10th century. It is a 'place-value' system, where the value of a particular digit depends on both the digit and its place in the number.

For example, the digit 2 represents very different values in the following numbers: 200 (2 x 100), 2000 (2 x 1000) and 1002 (1000 + 2). Place-value systems are more compact than 'additive' systems, such as Roman numerals, where, for example, 300 is represented by writing down the 100 symbol, C, three times: CCC.

Our number system is decimal – based on ten. But any number can be the base for a number system – for example, the sexagesimal system (base 60) used for time: 60 seconds = 1 minute, 60 minutes = 1 hour.

System	1	2	3	4	5	6	7	8	9	10	0
Mesopotamian Cuneiform marks for 1 and 10 only; a place-value system, base 60; no zero.											
Egyptian Hieroglyphs for 1, 10, 100, 1000, 10000, 100000, 1 million; an additive system, decimal (base 10); no zero.											
Mayan Three symbols; place-value with spaces between groups of symbols, to base 20.											
Greek Letters for 1–9, tens 10–90, and hundreds 100–900; additive; decimal; no zero.	A	B	Γ	Δ	E	F	Z	H	Θ	I	
Roman Seven symbols: I, 1; V, 5; X, 10; L, 50; C, 100; D, 500; M, 1000; additive-subtractive (IX = 9, XI = 11); decimal; no zero.	I	II	III	IV	V	VI	VII	VIII	IX	X	
Hindu A purely positional decimal system including a zero, requiring symbols for 0–9 only.											
Arabic/European Used in about the 15th century, adapted from the Hindu system.											
Modern Arabic/European The system now used universally.	1	2	3	4	5	6	7	8	9	10	0
Electronic binary Base 2. Used by electronic computers, because the two symbols equate to the basic electronic states 'on' and 'off'.	0001	0010	0011	0100	0101	0110	0111	1000	1001	1010	0000

BACKGROUND IMAGE A bill for the sale of a field and house written in cuneiform script (*c.*2550 BC).

Types of number

Natural (or whole) numbers – 1, 2, 3, 4 and so on – are the simplest form of number, used for counting whole units.

Integers include all natural numbers, together with zero and the negative forms of natural numbers: –4, –3, –2, –1, 0, 1, 2, 3, 4 and so on.

Rational numbers include all integers and also any other number that can be expressed as an exact **fraction** ($\frac{1}{3}$, $\frac{5}{8}$) or **decimal** (1.125, 0.003) – including numbers whose decimal expression ends in an infinitely recurring pattern, such as 0.33333.

Irrational numbers cannot be expressed as an exact fraction, though they still have precise values. The square roots of 2, 3 or 5 are examples, as is pi.

Real numbers comprise all rational and irrational numbers.

Imaginary numbers are those that are impossible by definition, such as the square root of a negative number (symbolised by i).

◆ **A factor** is a number that divides exactly into another number – the factors of 12 are 1, 2, 3, 4, 6 and 12.

◆ **A prime number** is a whole number that can be divided exactly only by itself and by 1, such as 2, 3, 53, 71.

◆ **A perfect number** is a whole number that is equal to the sum of all its other factors: 28 is equal to the sum of its factors 1, 2, 4, 7 and 14.

◆ **Infinity** (∞) is the symbol for the mathematical quantity that is larger than any possible number.

Large numbers

Large numbers generally have Latin names derived from 'million'. US usage is standard.

Billion	1 with 9 zeroes
Trillion	1 with 12 zeroes
Quadrillion	1 with 15 zeroes
Quintillion	1 with 18 zeroes
Sextillion	1 with 21 zeroes
Septillion	1 with 24 zeroes
Octillion	1 with 27 zeroes
Nonillion	1 with 30 zeroes
Decillion	1 with 33 zeroes

The term **googol** (1 with 100 zeroes) was coined by the nine-year-old nephew of the mathematician and author Edward Kasner.

769

What is pi?

Pi (π) is defined as the ratio of a circle's circumference to its diameter. It is used to calculate the length of curves and to make many other calculations. The fraction $\frac{22}{7}$ or 3.14159 is used as an approximate equivalent.

Sequences

Number sequences are sets of numbers in which successive numbers are derived according to a rule; for example, the sequence 1, 2, 3, 4, 5 … is formed by the rule 'add one to the previous number'.

In an **arithmetic sequence**, the difference between successive numbers is fixed (the **common difference**); for example, 2, 7, 12, 17, 22 … is an arithmetic sequence with a common difference of +5.

In a **geometric sequence**, the difference between successive numbers is determined by multiplying by a fixed amount (the **common multiple**); for example, 2, 4, 8, 16, 32 … is a sequence with a common multiple of 2

WEIRD AND WONDERFUL

The **Greek** and **Roman** number systems, which used new symbols for increasing power of 10, theoretically required an infinite number of numerical symbols. The **Mayan** system could express any number using just three symbols.

Fibonacci sequence

In the number sequence below each successive number is obtained by adding together the previous two numbers. The sequence was discovered by Leonardo Fibonacci in the 13th century, and has unexpected applications in art, architecture, astronomy and mathematics.

0   1   1   2   3   5   8   13   21   34   55

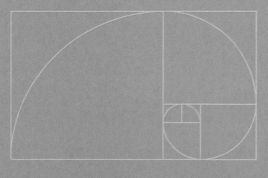

The classical geometric proportion known as the **Golden Section** is 1:1.618, which is almost identical to the proportions 8:13, 13:21, and so on found in the Fibonacci sequence. The Golden Section is thought to be naturally pleasing to the eye and is found in art and architecture.

FIBONACCI SPIRAL A series of rectangles can be contructed based on the Fibonacci sequence or the Golden Section proportions. A spiral drawn through the rectangles will have the same proportions as the spiral on a nautilus shell.

46

Using numbers

QUESTION NUMBER

The numbers or star following the answers refer to information boxes on the right.

ANSWERS

32	Fraction ❷
35	Percentage ❸
71	Algebra ❻
93	The lower number ❷
95	Calculus – in Latin, a small stone used in reckoning
99	One-third ❷
140	B: John Napier ❽
402	Two metres square – equals four square metres
434	Fermat ❽
510	A quarter ($\frac{1}{2}$ x $\frac{2}{3}$ x $\frac{3}{4}$ = $\frac{6}{24}$ = $\frac{1}{4}$) ❷
641	An abacus ❶
695	128 (dividing by $\frac{1}{4}$ is the same as multiplying by 4) ❷
★ 748	Andrew Wiles ★
827	C: Multiplication ❽
961	Subtraction – from Latin *subtrahere*, to draw away
968	Newton (Isaac) ❼ ❽

Core facts ❶

◆ The oldest mechanical aid to calculation is the **abacus**, still widely used in Asia and used in Europe until the 17th century.
◆ Leonardo of Pisa (also known as Leonardo Fibonacci) introduced **Arabic numerals** – 1, 2, 3, 4, 5, 6, 7, 8, 9 and 0 – to European mathematicians in the 13th century.

◆ The earliest reference to **decimals** – correctly known as decimal fractions – is by a 10th-century Arab mathematician, al-Uqlidisi. They were popularised in Europe in the late 16th century by Flemish mathematician Simon Stevin.
◆ From the 17th century, **calculus** made a new range of practical calculations possible.

Fractions ❷

In a fraction, the number above the dividing bar is the **numerator**, and the number below it is the **denominator**:

$$\frac{6}{8} \quad \frac{3}{4}$$ — numerator / denominator

A **proper** (or **vulgar**) **fraction** is one in which the numerator is smaller than the denominator, such as $\frac{5}{6}$. An **improper** fraction has a larger numerator, such as $\frac{7}{5}$.

If fractions have the same denominators – for example, $\frac{3}{7}$ and $\frac{2}{7}$ – **addition** and **subtraction** are carried out by adding or subtracting the numerators: $\frac{3}{7} + \frac{2}{7} = \frac{5}{7}$. If the denominators are different – for example, $\frac{3}{8}$ and $\frac{1}{3}$ – the fractions have to be converted into 'equivalent fractions' with the same denominator. For example, $\frac{3}{8} + \frac{1}{3} = \frac{9}{24} + \frac{8}{24} = \frac{17}{24}$.

To **multiply** fractions, multiply the numerators together and multiply the denominators together. For example, $\frac{2}{3}$ x $\frac{1}{5} = \frac{2}{15}$. To **divide** fractions, change the sign from ÷ to x, and turn the second fraction upside down. For example, $\frac{2}{3} ÷ \frac{5}{6} = \frac{2}{3}$ x $\frac{6}{5} = \frac{12}{15} = \frac{4}{5}$.

To **convert fractions to decimals**, simply divide the numerator by the denominator. For example, $\frac{3}{5} = 3 ÷ 5 = 0.6$.

Calculating percentages ❸

◆ A percentage is a fraction whose denominator is **100**. So 30 per cent = $\frac{30}{100} = \frac{3}{10} = 0.3$.

The use of percentages arose from medieval accounting's need for an easy basis for comparison. The symbol % was first used, as a shorthand, in Italy around 1425.

◆ To calculate a percentage of a number – say, 25 per cent of 10 – multiply the **first number** by the **second**, then divide by 100.
In this example, 25 x 10 = 250, 250 ÷ 100 = 2.5. So 25 per cent of 10 is 2.5.

◆ To calculate one number as a percentage of another – for example, 48 as a percentage of 75 – divide the **first number** by the **second**, then multiply by 100.
In this example, 48 ÷ 75 = 0.64, 0.64 x 100 = 64. So 48 is 64 per cent of 75.

◆ To calculate percentage change when an amount is increased or decreased – from 24 to 30, for example – divide the **difference** by the **starting quantity** and multiply by 100.
In this example, the difference between 24 and 30 is 6, 6 ÷ 24 = 0.25, 0.25 x 100 = 25. So the percentage change is 25 per cent.

Logarithms ❹

A logarithm is a way of expressing one number as the **exponent** (see 'Powers and roots', opposite) of another. For example, 1000 = 10 x 10 x 10 = 10^3. The exponent is 3, so the logarithm of 1000 to 'base' 10 is 3, or $\log_{10}1000 = 3$.

Before modern calculators, tables of 'logs' and 'antilogs' (converting numbers to logs and back again) were used to simplify **multiplication** and **division**, because adding logs is equivalent to multiplying the numbers they relate to.

For example, to **multiply 6 by 14** using logs, look up the logs for 6 and 14: $\log_{10}6$ = 0.7782, $\log_{10}14$ = 1.1461. Add them: 0.7782 + 1.1461 = 1.9243. Look up the antilog of 1.9243: 84. In other words: $6 \times 14 = 10^{(0.7782 + 1.1461)} = 10^{1.9243} = 84$.

Powers and roots

⑤

The term **power** indicates how many times a number has been multiplied by itself. So 9, for example, can be expressed as 3 to the power 2 or 3^2 – because 3 x 3 = 9. And 27 can be expressed as 3 to the power of 3 or 3^3 – because 3 x 3 x 3 = 27. The small raised number is known as the **index** or **exponent**.

Any number raised to the power of 2 is said to be **squared** – so 2 squared, or 2^2, is 4. Any number raised to the power of 3 is **cubed** – so 2 cubed, or 2^3, is 8.

Any number to the power of 1 is itself – $5^1 = 5$.

A **negative index** – say, 8^{-2} – shows how many times a number has to be divided into 1. For example, $8^2 = 64$, so $8^{-2} = 1 \div 64$ = $1/64$ or 0.0156. By the same principle, $10^{-1} = 1 \div 10^1 = 1/10$ or 0.1; $10^{-2} = 1 \div 10^2 = 1/100$ or 0.01; $10^{-3} = 1 \div 10^3 = 0.001$.

If a larger number is expressed as the power of a smaller one – for example, 16 as 4^2 – the smaller number is known as the **root** of the larger one. A root to the power of 2 is a **square root** – so 4 is the square root of 16. A root to the power of 3 is a **cube root** – $4^3 = 64$, so 4 is the cube root of 64. Beyond that come the fourth root, fifth root and so on.

Roots are indicated by the symbol $\sqrt{\ }$. For example, $\sqrt{64} = 8$ ($8^2 = 64$), $\sqrt[3]{64} = 4$ ($4^3 = 64$), $\sqrt[4]{64} = 2$ ($2^4 = 64$).

In science and mathematics, very large numbers are often expressed in a form known as **scientific notation** (or standard form). For example, the number 60 000 000 000 000 (60 trillion) can be expressed more simply as 6×10^{13} – in other words, 6 with 13 noughts after it.

Algebra

⑥

The word algebra comes from the Arabic *al-jebr*, 'the science of reuniting'. It involves finding **unknown numbers** from given information. For example, at its simplest, if x + 5 = 11, what is x? Subtract 5 from both sides of the equation: x = 11 – 5 = 6.

Basic algebra was studied from the 18th century BC in ancient Mesopotamia, Egypt, China and India. The Persian **al-Khwarizmi** developed the balancing process for solving equations in the 9th century AD. **François Viète** developed modern algebraic symbols in the 16th century. Nowadays, algebra is increasingly abstract, with applications in set theory, geometry, topography, physics and computing.

Calculus

⑦

Calculus is probably the most widely used branch of mathematics, relating to many real-world problems. It involves the study of **continuously changing variables** – for example, speed, stresses in engineering and electrical current – usually expressed as curving lines on graphs.

Its two branches are: **integral calculus**, which deals with the effects of continuously changing variables when added together; and **differential calculus**, which is concerned with rates of change.

The basic ideas of calculus have been studied since the times of the ancient Greeks, but were formulated in the late 1660s by **Leibniz** and **Newton**.

⭐ **748**

Last theorem

Pierre de Fermat's 'last theorem' was found after his death, scribbled in the margin of a textbook. It stated that the equation $x^n + y^n = z^n$ cannot be solved, where x, y, z and n are integers (see page 45) greater than 2. In 1993, **Andrew Wiles** of Princeton University claimed to have proved the theorem. He had to withdraw this initial claim, but put forward a revised proof the next year.

Mathematicians

⑧

◆ **Jacques Bernoulli** (1654-1705) Swiss mathematician. Together with his brother Jean (1667-1748) and nephew Daniel (1700-1782), he helped develop calculus and probability.

◆ **Georg Cantor** (1845-1918) German pioneer of set theory.

◆ **Pierre de Fermat** (1601-1665) French founder of probability theory and modern number theory, and author of the famous 'last theorem'.

◆ **Kurt Gödel** (1906-78) Austrian philosopher who proved that mathematics can never be totally consistent and totally complete.

◆ **David Hilbert** (1862-1943) German who tried to give mathematics a consistent and complete logical basis.

◆ **Muhammad ibn Musa al-Khwarizmi** (AD c.780-c.850) Persian founder of algebra, who also helped to develop the Hindu-Arabic decimal number system.

◆ **Gottfried Wilhelm Leibniz** (1646–1716) German philosopher who formulated calculus, independently of but concurrently with Newton.

◆ **John Napier** (1550-1617) Scottish inventor of logarithms and a calculating device, 'Napier's Bones'. Made the first use in a book of the decimal point.

◆ **John Forbes Nash** (1928-) American mathematician, whose work in games theory (the mathematical analysis of situations involving conflicts of interest) earned him the Nobel prize for economics in 1994.

◆ **Sir Isaac Newton** (1642-1727) English mathematician and physicist. His work included the formulation of calculus.

OSCAR NIGHT Nobel prize winner John Forbes Nash (far left) was portrayed by actor Russell Crowe (left) in the 2001 film *A Beautiful Mind*.

Geometry and trigonometry

Core facts ❶

◆ **Geometry** and **trigonometry** are related sciences concerned with length, angle, shape, area and volume.
◆ Geometry is the mathematical study of space, using **plane** (two-dimensional) and **solid** (three-dimensional) figures.
◆ Trigonometry is concerned with the properties of **triangles**. It is used in surveying, navigation and physics to calculate the heights and distances of inaccessible points.

Points and lines ❷

A **point** is a position in space; it has no dimensions (no length, width or depth).

A **line** connects two or more points, and has a single dimension (length); it may be straight or curved. Types of line include:

AXIS OF SYMMETRY A line dividing a shape into two reflecting (mirror) halves.

PERPENDICULAR A straight line that meets another at a right angle (an angle of 90°).

PARALLEL Lines that extend in the same direction and are a consistent distance apart.

TANGENT A line that meets a curve at one point only, without crossing it.

Angles ❸

An angle is the shape formed by two lines intersecting at a single point (known as the **vertex**). Angles are measured in **degrees** (°) or **radians** (rads). There are six basic types:

OBTUSE ANGLE An angle greater than 90° but less than 180°.

STRAIGHT ANGLE An angle of exactly 180°.

ACUTE ANGLE An angle measuring less than 90°.

RIGHT ANGLE An angle measuring exactly 90°.

REFLEX ANGLE An angle greater than 180° but less than 360°.

ROUND ANGLE An angle of exactly 360°.

Quadrilaterals ❹

A quadrilateral is a plane figure with four straight sides (a four-sided **polygon**). The four internal angles of a quadrilateral always add up to 360°.
  There are six basic types:

PARALLELOGRAM Opposite sides are parallel and equal in length. Opposite angles are equal.

RHOMBUS A parallelogram with four equal sides.

RECTANGLE Opposite sides are parallel and equal in length. The four internal angles each measure 90°.

SQUARE A rectangle with four equal sides.

TRAPEZIUM Two sides are parallel and unequal in length; two sides are not parallel and may be equal or unequal in length.

KITE Two pairs of adjacent sides are equal; opposite angles are equal.

GEOMETRIC CANOPY A stunning steel-lattice roof covers the Great Court at the British Museum.

Circles, spheres and cones 5

Geometry is also concerned with the properties of curves and the shapes they enclose, both **plane** (such as a circle) and **solid** (such as a sphere).

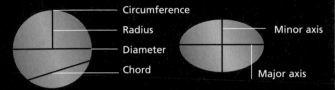

Circumference
Radius
Diameter
Chord
Minor axis
Major axis

CIRCLE A flat shape enclosed by a single curved line (the circumference), with all points on the line equally distant from the centre.

ELLIPSE An oval, technically defined as a cross-section of a cone (see below) that does not pass through the base of the cone and is not parallel to it.

CYLINDER A tubular solid shape that has straight sides and is circular in cross-section.

CONE A solid shape with a circular base, and sides that taper to a point at the apex (top).

Triangles 6

A triangle is a plane figure with three straight sides. The three internal angles of a triangle always add up to 180°. There are six basic types:

EQUILATERAL TRIANGLE Has three equal sides; internal angles each measure 60°. There are three axes of symmetry.

ISOSCELES TRIANGLE Has two equal sides and two equal internal angles; it has one axis of symmetry.

SCALENE TRIANGLE Has three unequal sides and three different internal angles; it has no axis of symmetry.

ACUTE-ANGLED TRIANGLE All three internal angles are acute (less than 90°).

RIGHT-ANGLED TRIANGLE Has one internal angle of exactly 90°. Other angles are acute.

OBTUSE-ANGLED TRIANGLE Has one internal angle greater than 90°.

How many sides? 7

Plane shapes enclosed by three or more straight sides are known as **polygons**. Apart from triangles and quadrilaterals, all are named from the Greek for the number of sides they have: **pentagon** (five sides), **hexagon** (six), **heptagon** (seven), **octagon** (eight), **nonagon** (nine), **decagon** (ten) and so on.

Trigonometry in practice 8

With a triangle, if you know the size of one acute angle (x) and the length of one of the sides, you can work out the length of the other two sides, using the **trigonometric ratios – sine** (sin), **cosine** (cos) and **tangent** (tan) – which can be looked up in tables or on a scientific calculator. For example, if x = 40° and the 'adjacent' side (a) = 4.5 cm, what is the 'opposite' side (o). According to the ratios, tan x = o ÷ a. Tan 40° = 0.839. So 0.839 = o ÷ 4.5. So o = 4.5 x 0.839 = 3.78.

h (hypotenuse)
o
x
a

Geometricians

◆ **Apollonius** (262-200 BC) The 'Great Geometer', who discovered and defined conic sections and the shapes they make.
◆ **René Descartes** (1596-1650) French philosopher who founded coordinate geometry.
◆ **Euclid** (c.330-260 BC) Greek mathematician whose work formed the basis of geometry for over 2000 years.
◆ **Leonhard Euler** (1707-83) Swiss mathematician who developed topology (the study of shapes and surfaces).
◆ **Nicolai Lobachevski** (1793-1856) Russian mathematician who suggested that

incompatible forms of geometry could 9 co-exist with Euclid's geometry.
◆ **August Ferdinand Möbius** (1790-1868) German mathematician and co-founder of topology, who invented the Möbius strip, a shape with only one edge and one side formed from a rectangle.
◆ **Pythagoras** (c.580-500 BC) Greek philosopher sometimes regarded as the founder of geometry.
◆ **Georg Friedrich Bernhard Riemann** (1826-66) German mathematician who developed non-Euclidian geometry.
◆ **Thales of Miletus** (c.624-c.547 BC) Greek philosopher who laid down five propositions on circles, lines and triangles.

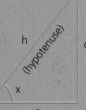

37

Pythagoras's theorem

The theorem provides a method for calculating the length of the longest side (the hypotenuse) of a right-angled triangle when the lengths of the other two sides are known. It states that in a right-angled triangle, the square of the hypotenuse equals the sum of the squares of the other two sides. Over 370 mathematical proofs of the theorem have been developed.

QUESTION NUMBER

The numbers or star following the answers refer to information boxes on the right.

ANSWERS

34 Average ❸

38 Probability ❶ ❷

147 True ❹

★ **243** Runners not listed have odds of 50/1 or more ★

451 *Casino* – directed by Martin Scorsese

529 50:50 (no difference how often flipped before) ❺

599 At a horse-race – signals used by bookmakers

631 D: Stake – from stake meaning post

632 C: Bingo – game can be traced back to 16th-c. Italy

633 B: Seven – 6 plus 1; 5 plus 2; and 4 plus 3

634 D: £1010 – you get the original stake back as well

636 D: 10 per cent – rate is measured per 1000 pop.

637 B: One in 35 – numbers are alternate red and black

638 A: Lotto (Ireland) ❺

639 B: Invented the slot machine – in 1895

640 B: Blaise Pascal ❶ ❷

693 Mensa – most intelligent 2% of the population

860 A lottery (run annually since 1812) ❺

969 Population – also called 'universe'

Probability and statistics

Core facts ❶

◆ Statistics and probability are related branches of mathematics concerned with patterns that occur in repeated but individually unpredictable events. They provide a means of analysing apparently random events and data collected from apparently very different sources.

◆ **Probability** is concerned with expressing mathematically the likelihood that a particular event will occur.

◆ **Probability theory** was initially developed in the 17th century by the mathematicians Blaise Pascal, Pierre de Fermat and Galileo Galilei.

◆ **Statistics** is the collection, interpretation and presentation of large amounts of numerical data.

Probability ❷

The probability of a **single event** occurring can be calculated by dividing the number of required outcomes by the total number of possible outcomes. The result is expressed as a fraction, a decimal fraction or a percentage.

Probability for **multiple events** depends on the relationship between them. If the events are **dependent** – that is, the result of one affects the probability of the other, as in throwing 5 then 6 in two dice throws – the probabilities are multiplied. Each outcome has a probability of one-sixth, but the second can only give the required result if the first has been successful, so the overall probability is 1/36.

If the required events are **independent**, as in throwing 5 or 6 in a single dice throw, the probabilities are added: a one-sixth chance of a 5 plus a one-sixth chance of a 6 gives a one-third chance of throwing a 5 or 6.

(Number of possible outcomes for 1 to 6 tosses of a coin)

```
                    1              1   (2)
                  1H-0T          0H-1T

              1          2          1   (4)
            2H-0T      1H-1T      0H-2T

         1         3         3         1   (8)
       3H-0T     2H-1T     1H-2T     0H-3T

      1       4        6        4       1   (16)
    4H-0T   3H-1T    2H-2T    1H-3T   0H-4T

   1      5       10       10       5      1   (32)
 5H-0T  4H-1T   3H-2T    2H-3T   1H-4T  0H-5T

  1     6       15      20       15      6     1   (64)
6H-0T 5H-1T   4H-2T   3H-3T    2H-4T  1H-5T  0H-6T
```

HEADS OR TAILS
The diagram shows the number of possible outcomes for tosses of a coin. From this the probability of a particular outcome can be calculated.

6 heads	5 heads 1 tail	4 heads 2 tails	3 heads 3 tails	2 heads 4 tails	1 head 5 tails	6 tails
1:64	6:64	15:64	20:64	15:64	6:64	1:64

Averages ❸

A useful way of making sense of numerical data is to find a typical, or average value. The three most important types of average are the mean, the median and the mode.

◆ The **mean**, or arithmetic mean, is calculated by adding the values together, then dividing the sum by the number of values. In the sequence below the mean is 40.09. In everyday usage, 'average' usually means arithmetic mean.

◆ The **median** value is the middle value in a set of numbers. In the example, which contains 11 numbers, 36 is the median value. If the number of values is even, the median is the arithmetic mean of the two middle values.

◆ The **mode** is the value that occurs most frequently in a list of values. In the example, the mode is 24, because this value occurs twice while the others occur once each.

| 24 | 35 | 68 | 25 | 46 | 79 | 24 | 45 | 36 | 38 | 21 |

| 21 | 24 | 24 | 25 | 35 | 36 | 38 | 45 | 46 | 68 | 79 |

Charts and graphs

4

Charts and graphs are frequently used to display statistical information visually. Such displays sometimes reveal patterns in data that can't be seen in a simple table.

◆ **Graphs** allow different kinds of data to be plotted against each other. A typical example might plot marks in an exam as a percentage along the horizontal axis (x-axis), and the total number of students achieving each mark – the **cumulative frequency** of the marks – along the vertical axis (y-axis).

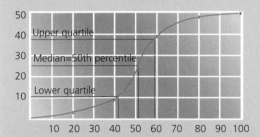

This type of display allows calculation of **percentiles**, or actual values, for analysing results. To find the 50th percentile, draw a a horizontal line from the mid point on the y-axis to the curve of results; a vertical line from that intersection to the x-axis shows the actual result at that percentile – 51 per cent in this example. The 25th percentile (the **lower**

quartile) and 75th percentile (**upper quartile**) are also often used for analysis.

◆ **Pie charts** (below) display statistical information by dividing a circle representing the total number of results into proportionally sized sectors.

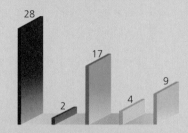

◆ **Bar charts** (above) show results as vertical or horizontal bars of proportionate length. A bar chart aids simple comparison of results, but can also reveal a **trend** in results if the bars are set out on a time axis (for example, monthly sales totals).

Betting

Gambling on sports is conducted using **odds**, a guide to the probability of a result expressed as the pay-out promised for a bet on it. A bookmaker's odds of 5/1 ('five to one') on a horse to win means that if the horse wins the race, the bookmaker will pay out five times the amount wagered (the **stake**) by the customer (less tax), plus the original stake.

WEIRD AND WONDERFUL

7

In reality, not all coins are 'fair' coins. Research has shown that Polish two zloty pieces, for instance, are more likely to land on heads than tails. The reason is that they are not symmetrical – the tails side is heavier, causing it to land face down.

One big lottery

5

Many countries use lotteries as a way of raising public money. The most famous is probably Spain's El Gordo ('The Fat One'), which has the **largest pool of prize money** in the world. Most lotteries use a fixed number of balls to give the winning numbers, so the probability of predicting the correct numbers can be calculated:

Lottery	Numbers/total balls	Odds per single selection
PowerBall (USA)	5/45 plus 1/42	1 in 80 089 128
The Big Game (USA)	5/50 plus 1/36	1 in 76 275 360
Lotto (Britain)	6/49	1 in 13 983 816
Pimera (Spain)	6/49	1 in 13 983 816
Lotto (Scandinavia)	6/48	1 in 12 271 512
Lotto (Hong Kong)	6/47	1 in 10 737 573
Tattslotto (Australia)	6/45	1 in 8 145 060
Lotto (Ireland)	6/42	1 in 5 245 786

Applied mathematics

6

Because they allow the mathematical manipulation of unknown, imprecise or apparently random events, statistics and probability have many modern applications.

◆ Business strategists often use **decision analysis**, or **game theory**, based on 'expected return' – the value

of a particular result weighed against the probability of its occurrence.

◆ Probability is used to analyse and predict the spread of diseases in the field of **epidemiology**.

◆ Many aspects of **atomic physics**, such as the distribution of electrons around a nucleus, are modelled

mathematically using probability.

◆ **Political** decisions – particularly the formulation of new social policies – are often based on detailed opinion polling and statistical analysis.

◆ Extensive use of sophisticated sampling techniques and statistics is a core part of **market research**.

Measuring time and distance

Core facts ❶

◆ Architecture and **surveying** in ancient Mesopotamia and Egypt first made it necessary to measure distance accurately.
◆ All early peoples used the stars as guides to position, but **true navigation** required accurate time-keeping as well as visual measurement.
◆ Early **sundials** kept 'temporary' hours that varied with the season. The first mechanical **clocks** varied by up to 1 hour a day; atomic clocks are accurate to 1 second per 1.7 million years.

TIMESCALE ❷

▶ **c.3500 BC** The ancient Egyptians make the first sundials.
▶ **c.1500 BC** The first known water clock is made, in New Kingdom Egypt.
▶ **AD 1090** The first recorded water clock fitted with an escapement is built, in China.
▶ **1353** The first known public clock is installed, in a tower in Milan, Italy.
▶ **c.1410** Architect Filippo Brunelleschi designs clocks with a spiral mainspring.
▶ **1510** Peter Henlein, a German locksmith, makes the first pocket watches.
▶ **1657** The first pendulum clock is made.
▶ **1671** English clockmaker William Clement invents the long (98 cm/39 in) pendulum.
▶ **1868** Swiss watchmaker Georges Frederic Rosskopf makes the first affordable pocket watch.
▶ **1922** English watch-maker John Harwood invents the self-winding mechanism.
▶ **1926** The Rolex Oyster, the first waterproof watch, is made in Switzerland.
▶ **1929** US clock-maker Warren Alvin Morrison invents the quartz crystal clock.
▶ **1971** The first digital watch, using a light-emitting diode (LED) display, is made by US engineers George Theiss and Willy Crabtree.

Running like clockwork ❸

Mechanical clocks combine a mechanism (rotating hands, for example) that makes regular movements with a means of counting these. The **escapement**, which regulates the movement and creates the clock's 'tick', was the main area of mechanical development.

TIME KEEPER Jim Gray – keeper of the American NBS atomic clock.

The world's first clocks ❹

The first **sundials** consisted of a simple pillar (gnomon) that cast a shadow on the ground; the earliest version with permanent reference marks dates from the 8th century BC, in Egypt. The Arab geometer Abu al-Hasan devised angled gnomons that eliminated seasonal variation in hour lengths in the early 13th century.

    Water clocks (clepsydras) were probably invented in ancient Mesopotamia; they measure either the falling level of water as it escapes from a container, or (a version from early America and Africa) the rate at which a floating vessel with a hole ships water and sinks.

PORTABLE TIMEKEEPER This ivory pocket sundial was made in Germany in 1592.

EARLY SEXTANT
The sextant is so named because the early instrument had a calibrated arc that was one-sixth of a circle.

Navigational devices

Navigation requires a means of calculating accurate positions on the Earth's surface. Calculations of **latitude** (distance north or south) have been possible since ancient times, by measuring the angle of the Pole Star to the horizon: the **astrolabe** (invented in the 6th century AD, though perhaps conceived in ancient Greece), **cross-staff** and **backstaff** (16th century), **quadrant** (1730) and **sextant** (1757) have allowed progressively more accurate observations.

Longitude (distance east or west) could be calculated only by reference to differing times as the Earth spins; accurate calculation became possible only in 1761, with John Harrison's **chronometer** No. 4, a highly accurate timekeeper for use on board ship.

The **compass**, allowing calculation of an accurate **bearing** (direction of travel) may have been invented in China as early as 2600 BC. The earliest European reference dates from 1217.

EARLY SEXTANT
The sextant is so named because the early instrument had a calibrated arc that was one-sixth of a circle.

TIMESCALE

▶ *c.*2300 BC The oldest surviving map is inscribed on a clay tablet in Mesopotamia.
▶ AD 100 Marinus of Tyre develops mathematical lines of longitude.
▶ *c.*150 Ptolemy of Alexandria creates the first great world map.
▶ 1730 John Hadley invents a 45° quadrant.
▶ 1757 John Campbell invents the 60° sextant.
▶ 1761 John Harrison's No. 4 chronometer sets new standards of accuracy.
▶ 1767 In Britain, *The Nautical Almanac* provides the first accurate tables showing the positions of stars and planets through the year.
▶ 1964 The US navy launches the first satellite navigation system.
▶ 1978–95 The USA sets up the Navstar global positioning system (GPS) to provide precise navigation.

Measuring distance

The Egyptian **cubit** (about 52 cm/21 in) was devised before 3000 BC; smaller units were the *djeba* (digit) at 28 per cubit, and the *shesep* (palm) of four *djeba*. Longer distances were measured using the *khet* (rod) of 100 cubits and the *iteru* (river) of 20000 cubits.

The Greeks divided the cubit into 24 *daktyloi* (digits) and introduced the **foot** of 16 *daktyloi*. The Romans subdivided the foot into 12 *unciae* (inches), and introduced the **pace** (about 1.5 m/5 ft) and the **mile** (1000 paces). Most of Europe used measurements partly based on the Roman system until the 19th century.

Metric measurements were first devised in Revolutionary France, and became the legal standard in 1840; they were adopted in most of Europe soon afterwards.

Calendars

The division of the year into 12 months based on lunar cycles was first made by the ancient Egyptians. The **Julian** calendar of the Alexandrian astronomer Sosigenes was introduced in 46 BC; it was based on the solar year of 365.25 days, and corrected the accumulated discrepancy of three months in the Egyptian calendar. The **Gregorian** calendar, which corrected a further accumulated discrepancy of ten days (about 11 minutes per year) in the Julian, was adopted in much of Europe in 1582 – but not until 1752 in Britain, and 1917 in Russia.

Barometer

⭐ 422

In 1643 the Italian physicist Evangelista Torricelli deduced **atmospheric air pressure** by observing that the level of mercury suspended in a glass tube rose and fell with changes in the weather. The aneroid barometer, which uses a vacuum chamber rather than mercury, was introduced in 1843.

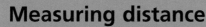

The mechanical world

Core facts ❶

◆ **Mechanics** is the part of physics that describes the effects of forces on objects. Its traditional subdivisions are **statics**, dealing with mass, weight and gravity, and **dynamics**, dealing with forces and objects in motion.
◆ Until 1900, mechanics was almost exclusively concerned with the behaviour of objects in the perceivable universe.

◆ The most influential physicist in traditional mechanics was **Sir Isaac Newton** (1642-77).
◆ Since 1900, mechanics has increasingly been concerned with the behaviour of objects beyond everyday perception, particularly those moving at near the speed of light.
◆ The key figure in modern mechanics was **Albert Einstein**, first theorist of **Relativity**.

Mass, weight and gravity ❷

Mass is the amount of matter that makes up an object. It is indistinguishable from weight in everyday life (and is measured in kilograms) – but is not determined by gravity.

Weight is the force exerted by an object when subject to gravity. In physics, weight is measured in **Newtons** (N) and is determined by multiplying the mass of an object by the force of gravity that is acting on it.

Gravity is the concept devised by Newton to explain the force that all objects exert on each other, proportional to their mass and distance apart. The Earth's gravity makes objects fall to the ground, and also keeps the Moon in orbit around it.

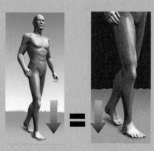

WEIGHT AND GRAVITY
A person's mass would have to be six times greater on the Moon to weigh the same as on Earth, where gravity is six times greater.

HOW FAR IN A SECOND

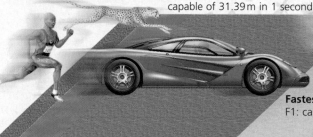

Fastest land animal The cheetah: capable of 31.39m in 1 second

Fastest production car McLaren F1: capable of 107.73m in 1 second

Fastest human American athlete, Tim Montgomery: 10.22m in 1 second on September 16, 2002

Force, motion and friction ❸

◆ A **force** is any influence that changes the position, shape, straight-line direction of motion or speed of an object.
◆ **Inertia** is the tendency of all objects to remain at rest or in motion unless affected by an external force.
◆ **Momentum** is the measure of the force needed to alter the motion of a moving object.
◆ **Friction** is the force that resists the relative movement of two objects when they are in contact.

Newton's laws ❹

Newton's three **Laws of Motion** (1687) are fundamental to traditional mechanics:
1. An object remains stationary, or continues moving in a straight line, unless an external force acts on it.
2. Any change in an object's position, motion or speed is proportional to the force acting upon it and to the object's mass.
3. Any action produced by a force is always accompanied by an equal and opposite reaction.

What is Relativity?

Einstein's **Special Theory** (1905) and **General Theory** (1915) of Relativity represent an entirely new attempt to explain motion, mass, energy, gravity and time.

The **Special Theory** addressed two discoveries of the 1880s: that the **speed of light** (written c in scientific formulae) is

$$E = mc^2 +$$

the only constant in a moving universe, and that Newtonian absolute motion and absolute space do not exist. It suggests that mass (m) and energy (E) are inter-convertible ($E = mc^2$), that the mass and length of an object must vary with its speed, and that time itself, within the frame of reference of an object moving near the speed of light, must slow down.

The **General Theory** extended these conclusions to propose that space and time are a single phenomenon, and that 'gravity' is in fact the result of **distortions of space-time** caused by mass. Though Newtonian gravity provides a workable explanation of the familiar universe, General Relativity also explains anomalies like the irregular orbit of Mercury, black holes and the bending of light by stars or planets.

TIMESCALE

▶ **330 BC** Greek philosopher Aristotle's *Physics* includes the first comprehensive study of mechanics.

▶ **AD 550** Johannes Philoponus suggests the velocity of a moving object is proportional to the excess of applied force over resistance.

▶ **1355** Jean Buridan develops a theory of impetus and air resistance to moving objects.

▶ **1440** Nicolas Cusanus theorises that the Earth is moving through space.

▶ **1604** Galileo Galilei proves that objects fall to Earth at a constant rate of acceleration, irrespective of mass.

▶ **1609** Johannes Kepler publishes his First and Second Laws of Planetary Motion.

▶ **1613** Galileo outlines the principle of inertia.

▶ **1687** Sir Isaac Newton publishes his Laws of Motion and Law of Gravitation.

▶ **1760** Mikhail Lomonosov publishes his laws of the conservation of energy and mass.

▶ **1797** Henry Cavendish makes the first accurate measurement of Newton's gravitational constant.

▶ **1887** Albert Michelson and Edward Morley prove the constant speed of light in a vacuum.

▶ **1889** George Fitzgerald proposes the 'Fitzgerald Contraction' (shortening in length) of objects as they approach the speed of light.

▶ **1905** Albert Einstein publishes his Theory of Special Relativity.

▶ **1915** Einstein publishes his Theory of General Relativity.

▶ **1919** Astronomical observation of the bending of light during an eclipse provides the first of many proofs of Einstein's theories.

Fastest land vehicle Thrust SSC: 341.11 m in 1 second on October 13, 1997

Fastest passenger plane Concorde: 597.22 m in 1 second at cruising speed

Going places

Scientific measurements are made using SI (*Système International*) units, based on the metric system.

◆ **Distance** is the length of a straight line between two places. It is a **scalar** quantity (one with magnitude but no direction).

◆ **Displacement** is a distance travelled plus the direction of movement. It is a **vector** quantity (one with a direction as well as magnitude).

◆ **Speed** is a scalar quantity, distance travelled divided by time expended – such as metres per second (m/s).

◆ **Velocity** is speed plus the direction of movement, a vector quantity.

◆ **Acceleration** is the rate of change in the velocity of a moving object, measured by dividing the change in velocity (m/s) by the time taken for this change – metres per second per second (m/s/s or m/s²). It is a vector quantity.

532

Einstein

The German-born Albert Einstein (1879-1955) was the greatest modern physicist, making breakthroughs in **radiation physics** and thermodynamics as well as Relativity. He received the **Nobel Prize for physics** in 1921, and had the newly discovered radioactive metallic element **Einsteinium** named after him in 1955.

GENIUS Einstein around the time he published his works on Relativity. He moved to America in 1933, after Hitler's rise to power in Germany.

Building & civil engineering 1

Core facts ❶

◆ Although they are closely related, **building** and **civil engineering** refer to different aspects of construction.
◆ **Building** refers to the construction of buildings. It began thousands of years ago when people first began making shelters.
◆ The design of buildings is the job of

architects. Structural **engineers** ensure that buildings are strong and stable.
◆ **Civil engineering** is the construction of public works such as roads, bridges, dams, pipelines and tunnels. The greatest ancient civil engineers were the Romans, who built road networks, aqueducts and drainage systems.

Architectural features ❷

Many key inventions have provided solutions to the problems of spanning openings and supporting roofs, allowing for larger and more complex interior spaces.

— Lintel

— Column or post

POST AND LINTEL A horizontal beam rests on vertical columns or posts. This is the simplest structure for creating doorways and supporting roofs.

— Keystone
— Round arch

ROUND ARCH The weight of the roof is transmitted directly downwards to supporting columns or walls. The Romans made extensive use of the round arch in their buildings.

— Pointed arch

GOTHIC ARCH Medieval architects developed the pointed arch. It could be used to span wider spaces than a semicircular arch, but buttresses were needed to resist the lateral (sideways) force exerted by the roof.

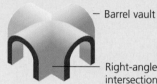

— Barrel vault

— Right-angled intersection

BARREL VAULTING Also used by the Romans, a round arch is extended to form a roof. Two vaults can meet at 90° to form a cross shape.

— Buttress

FLYING BUTTRESS These were a late Gothic invention. An arch carries the weight of the upper levels and roof to a buttress a short way out from the building.

— Spherical dome

DOME Yet another Roman invention, this is a kind of three-dimensional arch. It can be spherical (round), or pointed if based on a gothic arch. Two vaults can meet at 90° to form a cross shape.

Monuments to God ❸

Today's tallest buildings are monuments to Mammon, but until the late 19th century, all the biggest buildings were religious. The ancient **Egyptian pyramids** were the burial chambers of pharaohs (regarded as living gods), while the **stepped pyramids** of Central America were sites of ritual sacrifice. The world's biggest wooden building remains a **Buddhist temple** in Nara, Japan. Throughout most of the Christian era **cathedrals** were the largest and tallest buildings – until the Eiffel Tower was built in 1889.

World's tallest buildings ❹

People have always tried to build higher and higher. The eight buildings shown here are currently the world's tallest, measured against the highest free-standing structure in the world – the Canadian National (CN) Tower in Toronto, a broadcast and observation tower.

T & C Tower	**Empire State Building**	**Shun Hing Square**	**CITIC Plaza**	**Jin Mao Tower**	**Sears Tower**	**Petronas Towers**	**CN Tower**
Kaohsiung (Taiwan) 348 m (1142 ft)	New York, 381 m (1250 ft)	Shenzhen (China), 384 m (1260 ft)	Guangzhou (China), 391 m (1283 ft)	Shanghai, (China) 421 m (1381 ft)	Chicaco, (USA) 442 m (1450 ft)	Kuala Lumpur, 452 m (1483 ft)	Toronto, (Canada) 553 m (1815 ft)

(Chart scale: 100 m, 200 m, 300 m, 400 m)

The world's biggest dome ❺

London's Millennium Dome, completed in 1999, is the largest dome in the world, 320 m (1050 ft) in diameter and covering an area equal to 12 football pitches. It is made from Teflon-coated glass-fibre fabric supported by a web of steel cables from twelve 90 m (295 ft) steel masts.

The building with the greatest floor area is the Pentagon in Washington DC, with an area of 61.6 hectares (152 acres). By volume, the biggest is the Boeing aircraft building near Seattle, measuring more than 13 million m³ (472 million cu ft).

WEIRD AND WONDERFUL ❻

Architectural **follies** in the United States include replicas of the Leaning Tower of Pisa, in Niles, Illinois; a floating Taj Mahal, in Sausalito, California; and the grotto of Lourdes, in Emmitsburg, Maryland.

★ 348

Reaching for the skies

In traditional buildings, the walls support the upper floors. With more than a few floors, walls would have to be too thick to leave much usable space.

In **steel-framed** construction, first used in 1885, a strong, light frame bears the weight, leaving clear floor space. This technique, combined with the recently invented **passenger lift**, allowed the height of buildings to rise rapidly.

TIMESCALE ❼

► *c.*6000 BC Sun-dried bricks used in Middle East.
► *c.*2650-2500 BC Pyramids of Giza built.
► *c.*2200 BC First known bridge built in Babylon.
► *c.*1500 BC Post-and-lintel system first used in Egypt.
► 4th century BC Romans first use the rounded masonry arch. First paved Roman military roads.
► 3rd century BC First main sections of Great Wall of China.
► 2nd-1st century BC Romans build aqueducts.
► AD *c.*126 Concrete used to build the dome of the Pantheon in Rome.
► 12th century Gothic buildings, with pointed arches and flying buttresses.
► 1756 John Smeaton rediscovers hydraulic cement, first used by the Romans in 3rd century BC.
► 1779 First cast-iron arch bridge, across River Severn.
► 1825-43 First tunnel dug under River Thames.
► 1826 First iron suspension bridge, across Menai Strait, Wales.
► 1845 First reliable process to make Portland cement.
► *c.*1850 Reinforced concrete introduced.
► 1851 Crystal Palace built in London.
► 1852 Elisha Otis (USA) invents safety passenger lift.
► 1855 Bessemer converter increases steel production.
► 1869 Suez Canal opens.
► 1883 Brooklyn Bridge opens in New York.
► 1885 First steel-framed skyscraper built – the 10-storey Home Insurance Building in Chicago.
► 1889 Eiffel Tower built in Paris – the world's tallest structure at 324 m (1063 ft).
► 1892 Prestressed concrete invented.
► 1894 First escalator opens on Coney Island pier.
► 1931 Empire State Building built in New York.

The numbers or star following the answers refer to information boxes on the right.

ANSWERS

113	An aqueduct ❹
114	The 18th century (1779) ❻
115	Newcastle-upon-Tyne – the steel arch Tyne Bridge
116	Cantilever ❺
119	Denmark and Sweden – completed 2000
★ 142	False ★
184	Sydney Harbour Bridge (a steel arch bridge) ❺
186	Iron Bridge ❻
187	Akashi-Kaikyo Bridge ❺
265	False – second longest after Japan's Seikan tunnel
345	Camber ❶
394	B: Japan (links Honshu with Awaji-shima) ❺
398	B: St Lawrence Seaway (USA/Canada) ❹
399	B: Australia ★
400	D: Vakhsh, Tajikistan ❹
531	Netherlands – in a story by Mary Mapes Dodge, 1865
551	Italy and Switzerland – completed 1906
553	The Romans ❷
556	The Euphrates ❷

Building & civil engineering 2

Roads ❶

The **Romans** built the first proper road network – to be able to deploy infantry rapidly. The **Industrial Revolution** brought major advances: around 1800, engineers such as Pierre Trésaguet in France and Thomas Telford and John McAdam in Britain discovered how to make rain-proof roads with layers of compacted, graded stones and a camber (curved top) for drainage. Compacted broken stones made the best surface, especially with tar added to make **'tarmacadam'**.

Much the same system is used today, with the addition of concrete.

Going underground ❷

The first known tunnel for pedestrians rather than mining was dug under the Euphrates river in Mesopotamia about 2100 BC. The **oldest-known road tunnel**, 38 m (125 ft) long, was dug by the Romans on the Via Flaminia in AD 76.

Explosives have been used for tunnelling in rock since the 17th century. In 1815, Marc Brunel invented the **tunnelling shield**, a protective tubular metal structure that is gradually pushed forwards as the workers hack away at the earth or rockface. Nowadays, most digging is done by huge drill-like tunnel-boring machines. Shallow underwater tunnels may be prefabricated in sections, then sunk to the bottom.

Some of today's record-holding tunnels are:
Longest rail tunnel Seikan (Japan, 1988), 54 km (34 miles)
Longest underwater tunnel Channel Tunnel (Britain/France, 1994), 50 km (31 miles), with 38 km (23 miles) under sea
Longest road tunnel Laerdal (Norway, 2000), 24.5 km (15 miles)
Longest (and oldest) metro system London Underground (begun 1863), 390 km (242 miles).

CHANNEL TUNNEL
On average the tunnel is 46 m (150 ft) deep under the seabed. There are three tunnels in all: two for trains and one for service.

Tie-breaker ❸

Q: What is the world's longest road?
A: The Pan-American Highway. It runs more than 24000 km (15000 miles) from Alaska to Chile, across to Argentina and north again to Brazil. The only gap is the Darién Gap in Panama and Colombia.

Waterways and waterworks ❹

The ancient Egyptians and Mesopotamians built dams and irrigation canals from about 3000 BC. The **longest canal** still in use is China's 1780 km (1110 mile) Grand Canal, started in 485 BC. By the 1st century AD, the Chinese were also building primitive **locks**, allowing boats to pass from one level of a waterway to another. The **longest canal system** for ocean-going vessels is North America's **St Lawrence Seaway**, 1200 km (745 miles) of rivers and canals, linking the Great Lakes with the Atlantic.

The ancient Romans had the biggest system of **aqueducts** for supplying water to their cities: they totalled about 610 km (380 miles). The longest aqueduct today is 1670 km (1040 miles) long, and supplies water to Tripoli in Libya. The **longest dam** is the 70 km (40 mile) Yaciretá-Apipé dam on the Paraná river between Argentina and Paraguay. The **tallest** is still under construction: the 335 m (1099 ft) high Rogun dam on the Vakhsh river in Tajikistan.

BIG WHEEL
The world's first rotating boat lift, opened at Falkirk in 2001, links Scotland's Forth & Clyde and Union canals.

Bridging the gap ⑤

◆ **Cantilever** bridges depend on the stiffness of the cantilever beam, which is supported at the centre. One end is anchored to the shore, the other is linked to a second cantilever.

◆ In **arch** bridges the shape of the arch itself supports the weight.

◆ In a **suspension** bridge, vertical supporting hangers link the deck to heavy suspension cables which pass over the towers and are anchored to the ground. The cables and towers bear the deck's weight.

◆ A **cable-stayed** bridge looks similar, but spaced supporting cables link the deck directly to the towers.

Longest cantilever Pont de Québec, Canada: 549 m (1800 ft).

Longest steel arch New River Gorge Bridge, West Virginia: 518 m (1700 ft).

Longest suspension Akashi-Kaikyo Bridge, Japan: 1990 m (6529 ft)

Longest cable-stayed Tatara Bridge, Japan: 890 m (2920 ft).

OLD AND NEW ⑥

The Iron Bridge (right) near Telford, in Shropshire, England, is the world's oldest cast-iron arch bridge, built in 1779. Below: The Millennium Bridge across the River Tyne in Newcastle was opened in 2001. It is a foot and cycle bridge, whose curved walkway can be swivelled upwards so that ships can pass beneath.

★ 142

Dividing lines

The world's **longest structure** is the Great Wall of China, about 6325 km (3930 miles) long. Contrary to popular belief, it is not visible from the Moon. The **longest fence** is Australia's Dog or Dingo Fence, stretching 8500 km (5300 miles) across Queensland and South Australia. It was built to keep dingoes out of sheep-grazing country.

Energy

Core facts ❶

◆ Everything that happens involves the **conversion of energy**. Most machines convert energy from one form into another.
◆ In science, **energy** is the capacity to do work.
◆ **Work** is what energy makes happen – a movement, an increase in temperature, or some other change. Energy and work are two sides of the same coin, and are measured in the same units: **joules** (J), **kilojoules** (kJ), **calories** (cal), and, for gas and electricity, kilowatt-hours (kWh).
◆ Energy cannot be created or destroyed, except by annihilating or creating matter.
◆ **Power** is a measure of how fast work is done, or how fast energy is converted or 'consumed'. It is measured in **watts** (W; joules per second), **kilowatts** (kW) or **horsepower**.

Temperature ❷

Temperature, the indication of an object's heat energy, is measured in degrees.

100 billion°C	Supernova explosion
1 billion°C	Nuclear explosion
510 million°C	Highest temperature ever achieved in a laboratory
14 million°C	Centre of the Sun
50 000°C	'Blue dwarf' stars
6000°C	Surface of the Sun
3380°C	Melting point of tungsten (highest for any metal)
2550°C	Light bulb filament
600°C	Molten lava
480°C	Surface of Venus
100°C	Water boils
57.8°C	Highest recorded on the surface of the Earth (Libya, 1922)
37°C	Normal body temperature
0°C	Water freezes
−18°C	Seawater freezes (0°F)
−38.9°C	Mercury freezes
−89.2°C	Lowest recorded on Earth (Vostok, 1983)
−141°C	Highest for superconductivity
−200°C	Air turns to liquid
−268.9°C	Helium turns to liquid
−270°C	Outer space
−271.4°C	Liquid helium freezes
−273.16°C	Absolute zero

Loudness ❸

A sound's loudness depends on its density, and is measured in decibels.

0 dB	Threshold of hearing (faintest audible sound)
10 dB	Rustling leaves; pin dropping 10 m (30 ft) away
20 dB	Soft whisper
40-50 dB	Normal conversation
60-70 dB	Street traffic
80 dB	Vacuum cleaner; factory; home hi-fi
90 dB	Heavy lorry traffic
98 dB	Symphony orchestra
100 dB	Personal stereo
110 dB	Rock concert (front row)
120 dB	Car 'boom-box' stereo
120–140 dB	Jet aircraft taking off
130 dB	Threshold of ear pain
140 dB	Gunshot
140-190 dB	Space rocket lift-off
160 dB	Volume at which eardrum ruptures

Energy transfer ❹

Imagine a man getting up. He switches on the light, converting electricity into heat and light. The man's breakfast contains chemical energy made by photosynthesis, the process plants use to convert sunlight (made by the release of nuclear energy) into food. Going for a run (doing work), he converts food energy into kinetic energy, body heat energy and, climbing a hill, potential energy.

During all this activity, no energy is created or destroyed. Energy is merely converted into or from different forms of energy, or into or from work. This is the law of **conservation of energy**.

CHEMICAL ENERGY
This is the energy **stored in chemical molecules**, such as those of the fuel in the Space Shuttle's tanks. When these burn, chemical energy is converted into heat energy. The products of this combustion shoot out of the rear of the rocket at high speed – they have kinetic energy. This in turn is converted into the kinetic energy of the Shuttle's movement, the potential energy of its height, plus sound and light energy.

ELECTRICAL ENERGY
Electricity, or electrical energy, is the movement of tiny, electrically charged subatomic particles called **electrons** (see page 62). When an electric current flows through a conductor such as a wire, electrons jump from atom to atom along the conductor. Electricity is easy to transport in this way, and is easily converted into other forms of energy to power devices such as motors, heaters and lamps. Virtually every Shuttle system, from fuel pumps to control-panel lights, uses electricity.

POTENTIAL ENERGY The higher the Shuttle rises against Earth's gravity the greater is its potential energy – the energy an object has because of its **position or shape**. As it returns to the ground, its potential energy is converted into kinetic energy (its speed) and the heat of re-entry.

NUCLEAR ENERGY This is the energy that **holds together** the nuclei of atoms (see page 114). Some of this energy is released in nuclear reactions, such as those powering the Sun and those in nuclear bombs and reactors. At the same time, matter is destroyed. It is converted into energy: mainly heat, the kinetic energy of particles formed in the nuclear reaction, and electromagnetic radiation such as gamma rays and light.

KINETIC ENERGY Any moving object has kinetic energy – the **energy of movement**. For two objects travelling at the same speed, the kinetic energy is proportional to mass – a 10 tonne truck has ten times the energy of a 1 tonne car. If speed doubles, energy is increased four times; if speed triples, energy is increased ninefold, and so on.

SOUND ENERGY This takes the form of **pressure waves** or vibrations – rapidly alternating high and low pressure – in the air, or in a solid or liquid medium (sound cannot travel through a vacuum). It is created by vibration of the hot gases shooting from the Space Shuttle's engines; in your ears it is converted into electrical impulses carrying messages to the brain.

HEAT ENERGY This is one of the most familiar forms of energy, and is simply what makes something hot. It is caused by the **vibration or movement of atoms** – the faster they move the hotter an object is. Only at absolute zero (impossible to reach) would all movement, and so heat, cease. Burning fuel creates heat in the engines of the space shuttle. Heat moves by **radiation** (like light); by **conduction** through a material (like electricity); or by **convection** (the flow of a hot liquid or gas upwards because it is lighter than cooler material surrounding it).

LIGHT ENERGY This is the best-known form of **electromagnetic energy** (see page 86). It travels at a constant, fixed speed through a vacuum, and slightly more slowly through other transparent media. It is produced by hot objects (such as the Shuttle's exhaust), by certain chemical and nuclear reactions, by fluorescence, or by other means.

GOING UP The Space Shuttle *Columbia* lifts off from the Kennedy Space Centre, 1996.

Energy types ❺
There are **several different types** of energy. All inert objects possess energy in some form or another. When objects move or alter in some way, work is done and energy is usually converted into other forms. This is illustrated by a Space Shuttle launch.

★ **544**

Sonic boom
A sonic boom is created whenever an object travels **faster than sound**. It is caused by air pressure waves piling up in front of the object (in the same way as a bow wave forms before a boat), then being released as a shock wave. The larger the object, the louder the boom. The **speed of sound** in air is 1220 km/h (758 mph). In denser media, such as water, it is faster.

Electricity and magnetism

QUESTION NUMBER

The numbers or star following the answers refer to information boxes on the right.

Core facts ❶

◆ Electricity and magnetism are closely linked. Both are caused by tiny, electrically charged subatomic particles called **electrons**.
◆ **Static electricity** normally stays put. It occurs where a build-up of electrons gives an object an overall electric charge.
◆ **Current** electricity flows through materials called **conductors** but not through **insulators**.
◆ **Magnetism** in a solid is caused by electrons spinning within the solid's molecules.
◆ Magnetism can also be caused by an electric current in a wire. If the wire is wound into a coil, the **magnetic field** – the area of magnetic influence – is stronger.

What is electricity? ❷

◆ Subatomic particles inside atoms have either a positive or negative **electric charge**. Normally, the negative and positive charges neutralise each other. But in some materials you can cause a surplus or deficit of electrons (negatively charged particles) by rubbing or other means. This creates **static electricity**.

◆ In conductors, the electrons on the outside of atoms are only loosely attached. When an **electric current flows** in a wire, the electrons jump from atom to atom along it. Each electron moves only a tiny distance, but a domino effect means they all move. The more electrons that pass a given point, the bigger is the current.

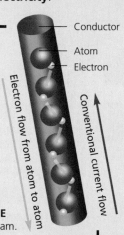

ONE WAY Electrons flow from the negative terminal of a battery to the positive. This is opposite to the arbitrary 'conventional' current, which is said to flow from positive to negative.

–

Conductor
Atom
Electron

Electron flow from atom to atom

Conventional current flow

BACKGROUND IMAGE An electric circuit diagram.

+

Lamp

Switch

ROUNDABOUT Electric current can only flow in a complete circuit, from the current source (such as a battery) through the various components and back to the source. Breaking the circuit anywhere (for example, with a switch) stops all current flowing.

Battery

Electricity in nature ❸

A **lightning** flash is caused by an imbalance in the positive and negative charges within a thundercloud, due to swirling ice particles. The top builds up a positive charge, the bottom a negative one. The difference can become so great that **static electricity** is discharged within the cloud and also between the cloud and the ground.
Much smaller voltages occur in living things. **Nerve impulses** consist of electrical discharges along nerves, and applying electricity causes contraction of muscles.

LIGHT SHOW Lightning is a natural discharge of built-up static electricity.

Conductivity ❹

Materials are classified according to how well they conduct electricity:
◆ **Conductors** have plenty of free (or loose) electrons, and offer little resistance to electric current.
◆ **Insulators**, such as rubber, have few free electrons. They have a high resistance to the flow of electricity.
◆ **Semiconductors** can be made to conduct or not depending on how voltage is applied. They are used to make microchips.
◆ **Superconductors**, such as metals at a temperature just above absolute zero, offer no resistance to current. Once it starts, it keeps flowing.

What is magnetism?
❺

◆ Magnetism **cannot be seen** except by its effect on strongly magnetic materials or electrical conductors.
◆ Spinning electrons in the atoms of these materials create force fields called **magnetic fields**. If these are aligned they reinforce each other and the material is a magnet.
◆ Each magnet has two ends, a north and a south pole. As with electric charges, like poles **repel** each other and opposite poles **attract**.
◆ The Earth acts like a magnet. Its **poles** are near the north and south geographic poles.

Electromagnetism
❻

Danish scientist **Hans Oersted** was the first to discover the link between electricity and magnetism. In 1819, he found that an electric current flowing in a wire makes a nearby compass needle move: the electric current creates a **magnetic field** around the wire. It was soon discovered that if the wire is wound into a coil, the coil reinforces the field. In 1825, **William Sturgeon** found that putting a soft iron core inside the coil increases the magnetic field's strength still further.

   Michael Faraday predicted the reverse effect – that a magnetic field generates an electric current. In 1831, he proved that when a magnetic field moves over a coil, a current is induced in it. This is the principle of the **electric generator**, which made possible the widescale generation and use of electric power. Faraday also discovered that a conductor carrying a current in a magnetic field experiences a force – the principle of the **electric motor** and the basis for a vast range of modern electric machines.

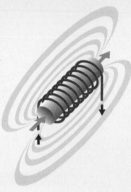

FORCE FIELD Current in a coil of wire (left) creates a magnetic field. When a generator's coil rotates between the poles of a magnet (below), an electric current is induced in it.

Superconductors
Close to absolute zero (-273°C), many metals become 'superconductors'. Once a current starts flowing, it continues **without losing strength** or heating the conductor. Some ceramic materials take on superconducting properties more than 100°C above absolute zero. This makes the **industrial use** of superconductivity possible.

WEIRD AND WONDERFUL **❼**
A magnet's north pole is the end that points towards north. But only **opposite magnetic poles attract** each other. This means that the Earth's North Pole is actually a magnetic south pole.

Michael Faraday
❽

Michael Faraday (1791-1867) had very little formal education but eventually became director of the **Royal Institution**, one of Britain's most important scientific foundations. He began work as a bookbinder at the age of 14. Inspired by books he bound, he attended lectures given by the eminent chemist Sir **Humphry Davy** at the Royal Institution, and – largely by chance – was offered a job as Davy's assistant. He soon began making his own discoveries. Working at first as a chemist, he discovered the petroleum-derived liquid **benzene** in 1825. But his greatest discoveries were in **electrolysis** (chemical changes caused by electric current) and **electromagnetism**.

FATHER OF ELECTRICITY Faraday photographed around the age of 70.

What's in a name?
❾

◆ **Ampère** or **amp** Unit of electric current. Named after French physicist **André-Marie Ampère** (1775-1836), who found how to detect and measure electric current.
◆ **Coulomb** Unit of electric charge. Named after French physicist **Charles-Augustin de Coulomb** (1736-1806), who experimented with electrostatic charges.
◆ **Farad** Unit of capacitance – the ratio of electric charge to potential on an electrically charged conductor. Named after **Michael Faraday**.
◆ **Ohm** Unit of resistance. Named after the German physicist **Georg Ohm** (1787-1854), who discovered the relationship between current, voltage and resistance.
◆ **Volt** Unit of electromotive force (emf) or voltage. Named after Italian physicist **Alessandro Volta** (1745-1827), who invented the first electric battery.
◆ **Watt** Unit of power – the rate energy is consumed or expended. Named after Scottish engineer **James Watt** (1736-1819), who developed the steam engine.

Uses of electricity

Core facts ❶

◆ Nothing changed people's lives so much in the 20th century as the **availability of electricity** and the invention of devices powered by it.
◆ The key inventions, after the basic electric motor and electric lighting, were those of **Nikola Tesla** for the generation, transmission and use of alternating current (AC). These enabled electricity to be **generated** efficiently at power stations, transmitted wherever it was needed and **used safely** to operate machines and other devices.
◆ A second, linked revolution was that of **electronics** (see page 108). This started with John Ambrose Fleming's invention of the diode valve in 1904, and was given a huge boost by the invention of the transistor in 1947 and the microchip in 1958.

Electric motors

The most important invention for putting electricity to work was the electric motor.

At its simplest, this consists of a **coil of wire** pivoted in the **magnetic field** between the poles of a permanent magnet or an electromagnet (known as the field magnet). Current flowing through the coil (known as the armature) creates its own magnetic field, with forces of **attraction** and **repulsion** between it and the magnetic field around it. The interaction of these forces creates movement: **the coil or armature spins**, driving a shaft.

A simple **direct current (DC) motor** operates on direct current (see page 67), usually from a battery. Power is fed to the armature through a pair of springy metal strips, known as **brushes**, one on each side of the armature. Another pair of plates, known as the **commutator**, is attached to the armature. Contact between the brushes and the commutator ensures a continuous rotary motion.

Other kinds of electric motor include:
◆ the **alternating current (AC) induction** motor, which operates only on alternating current (see page 67). The armature consists simply of a series of thick copper loops embedded in the surface of an iron core and joined at each end to a copper ring.
◆ the **AC synchronous** motor, which always turns at a constant speed, depending on the frequency of the AC supply. It is used in electric clocks. ❷

HOT HEADS Advertising the Pifco Princess in 1963.

Home comforts

The heating effect of an electric current, discovered by British scientist Sir **Humphry Davy** in 1801, is responsible for most domestic electric heating and lighting. The first device to use the principle was the **arc lamp**, invented by Davy in 1807. Many people tried to develop a light with a heated filament enclosed in a glass bulb. The first to succeed were the Englishman **Joseph Swan** and the American **Thomas Edison**, both in 1878-79. Later improvements included the **tungsten filament** (1907), and the use of inert gases such as argon to increase bulb life (1913).

In 1902, Georges Claude discovered the lighting effect of passing electricity through **neon**. By the 1920s multicoloured tubular signs (using different gases for different colours) had spread around the world. In the 1930s the same principle was extended to produce **fluorescent strip-lighting**.

Domestic **electric heating** was first introduced in the 1880s. The heater element of the basic 'electric fire' glows red-hot, and gives heat both by radiating infra-red (see page 86) and by heating the air in contact with it.

❸ **STYLE LIGHT** The horn-shaped 'Ora' lamp, created by the Parisian designer Philippe Starck.

TIMESCALE

▶ **1790s** Italian Alessandro Volta invents first useful electric cell.
▶ **1829** American Joseph Henry makes first useful electric motor.
▶ **1831** Faraday invents electric generator and rotating motor.
▶ **1840** First electric clock.
▶ **1843** Scotsman Alexander Bain patents the first fax machine.
▶ **1860s** Frenchman Georges Leclanché invents cell from which modern dry-cell battery evolved.
▶ ***c.*1873** German Karl von Linde introduces compression refrigerator.

PORTABLE WARMTH
An early electric fire.

▶ **1877** Thomas Edison invents the phonograph.
▶ **1881** Alexander Graham Bell, inventor of the telephone, invents the metal detector.
▶ **1888** Croatian-American Nikola Tesla patents AC electricity generator, motor and transformer – the basis for the widespread use of electric power.
▶ **1889** First electric oven.
▶ **1890** First electric train.
▶ **1893** First electric toaster.

▶ **1901** British engineer Hubert Booth invents the vacuum cleaner. ❹
▶ **1902** First air-conditioner.
▶ **1904** In Britain, John Ambrose Fleming patents the vacuum diode valve – the first electronic device.
▶ **1907** Long-life tungsten-filament electric light bulb introduced. First electric washing machine.
▶ **1910** Electric food mixer patented.
▶ **1923** First efficient electric kettle.
▶ **1926** First electric lawnmower.
▶ **1927** American Philo Farnsworth makes the first purely electronic television transmission.
▶ **1931** First electric razor.
▶ **1935** American Adolph Rickenbacker invents the electric guitar.
▶ **1938** American Chester Carlson invents electrostatic photocopying.
▶ **1946** Microwave oven invented.
▶ **1947** Americans John Bardeen, Walter Brattain and William Shockley invent the transistor.
▶ **1948** First programmable general-purpose computer built at Manchester University.
▶ **1953** First transistor radio.
▶ **1958** American Jack Kilby builds first integrated circuit (microchip).
▶ **1960** First cardiac pacemaker – the size of a TV set.
▶ **1979** Akio Morita, head of Sony, invents the personal stereo.

CLEANING UP English inventor James Dyson unveiled his two-drum washing machine in 2000.

Thomas Edison ❻

The 'father of invention', Thomas Alva Edison (1847-1931), registered 1093 US patents – more than any other individual. A systematic inventor, he set up an **'inventions factory'** in 1876, where he and his assistants worked to develop products that could be put into production.

Edison's first patent was for an electrical vote-indicator, but many of his other early inventions were for telegraphy. He tinkered with the **telephone**, and a year after Alexander Graham Bell's success developed an improved mouthpiece that was used for 100 years. His best-known inventions were the **phonograph** (gramophone) and (with Joseph Swan) the **electric light bulb**.

WEIRD AND WONDERFUL ❼

Many inventions have significance that is not realised for years. When **Michael Faraday** was asked what was the use of his electricity-generating dynamo, he could not think of one, but replied: 'What use is a baby?'

 705

Diode valve

A diode is a device that allows current to flow in one direction only, and is used as a 'rectifier' to turn alternating into direct current. English engineer **John Ambrose Fleming**'s diode valve, patented in 1904, opened up the development of radio (by turning AC radio signals into DC signals that can be picked up by a receiver) and electronics.

Generating energy 1

Core facts ❶

◆ Almost 90 per cent of the world's total energy consumption (and nearly 65 per cent of its electricity) comes from burning **fossil fuels** – oil, coal or natural gas.
◆ **Nuclear power** accounts for under 8 per cent of world energy consumption.
◆ **Hydroelectric plants** provide about 2.5 per cent of the world's energy, and a tiny amount comes from other **renewable** sources.
◆ The USA consumes a quarter of all the world's energy.
◆ Iceland, Malawi, Norway, Paraguay and Zambia generate more than 99 per cent of their electricity from **renewable** sources.

How a power station works ❷

Coal-fired power stations produce steam to **drive turbines** and **generate electricity**. Coal is ground up into small pieces to make it burn faster, and is mixed with pre-heated air before being burnt inside a furnace. As the coal burns, the heat is used to boil water contained in pipes. The water becomes **super-heated steam**, which reaches a very high pressure. The steam is forced through the turbines, which turn an alternator (alternating-current generator) to make electricity.

Once the steam has been passed through the turbines, it is cooled by a condenser. The resulting water is returned to the furnace for reheating.

COAL-FIRED Coal is still an important energy source, which is predicted to last for another 40 years.

The hot flue gases pass through filters to remove dust and other harmful emissions. They are also used to preheat the ground-up coal.

High-pressure steam is forced into the turbines.

Cold water is delivered to the condenser to cool the steam from the turbines.

Coal is ground into small pieces.

The coal is burnt in the furnace, producing steam.

Steam hits the fans of the turbines and makes them spin.

Hot water in the condenser (heated by the steam leaving the turbines) is sent to cooling towers to be cooled and reused.

The alternator driven by the steam turbines generates electricity.

★ **845**

Hot rocks

Iceland lies on a spot where the Earth's crust is spreading. Hot, molten rocks, known as magma, well up near to the surface, sometimes breaking through the ground to form lava flows. Water seeping (or pumped) down into the hot rocks is released as **geothermal steam** to generate electricity, or as **hot water** to warm Reykjavik's buildings. Iceland generates the rest of its electricity in hydroelectric plants.

Power battle ❸

Although the American inventor Thomas Edison built the world's first proper power station in 1882, he backed the wrong invention in this case. His plant produced direct current (**DC**) electricity. His rival **George Westinghouse**, using the inventions of Croatian immigrant **Nikola Tesla**, promoted alternating current (**AC**).

The advantage of AC is that its voltage can be stepped up or down with a transformer. High-voltage power can be transmitted over long distances with little loss of power, then stepped down for safe use. DC cannot be transformed in this way, and transmission over long distances involves great losses from heating the cable.

Today, virtually all the world's generating plants use Tesla's and Westinghouse's system.

POWER PLANT Red aircraft warning lights identify the chimney stacks of a coal-fired power station near Cologne in Germany.

TIMESCALE ❹

► **370 BC** The first known use of coal as fuel, in China.

► **AD 915** The earliest known use of windmills, to grind grain in Persia.

► **1765** Coal gas is first used for lighting, in mine offices in Cumberland.

► **1807** Gas street lighting is first used, in London.

► **1815** English inventor Humphry Davy invents the miner's safety lamp.

► **1821** The first natural gas well is dug at Fredonia, New York State.

► **1855** Robert Bunsen invents a gas burner.

► **1859** The first oil well is drilled, at Oil Creek, Pennsylvania.

► **1872** The first natural gas pipelines are built in New York and Pennsylvania.

► **1881** An experimental hydroelectric power plant is built in Germany. Godalming, Surrey, becomes the first British town with electricity generated by water-wheel.

► **1882** Thomas Edison opens the world's first full-scale electricity generating plant in New York.

► **1893** The opening of the world's first full-scale hydroelectric power station at Niagara Falls.

► **1904** The world's first geothermal power plant opens, at Lardarello, Italy.

► **1938** Otto Hahn and Fritz Strassmann in Germany split uranium atoms with neutrons; this releases more neutrons, showing the possibility of nuclear chain reactions.

► **1941** The world's first modern wind generator is built in Vermont.

► **1942** The first nuclear reactor in operation at the University of Chicago.

► **1948** The discovery of the Al-Ghawar oil field in Saudi Arabia, later proved to be the world's biggest.

► **1952** The first experimental fast-breeder nuclear reactor opens in Idaho.

► **1954** The world's first nuclear power plant opens at Obninsk in the Soviet Union.

► **1960** The first solar power plant is built in Turkmenistan.

► **1967** The first major tidal power plant is built in the Rance estuary, France.

► **1979** The partial meltdown of the Three Mile Island nuclear power plant, Pennsylvania.

► **1982** The Solar One solar power plant in operation in the Mojave Desert.

► **1986** A conventional explosion in a reactor at Chernobyl, Ukraine, causes widespread nuclear pollution.

► **1991** Scientists at the Joint European Torus facility in Oxford achieve a controlled nuclear fusion.

Fossil fuels: pros and cons ❺

Fossil fuels – **coal**, **oil** and **natural gas** – were formed millions of years ago from the remains of plants and animals that were compressed and chemically changed as rocks formed on top of them.

Pros: Fossil fuels can be extracted – and burned to generate heat and electricity – quite easily and cheaply. Oil and gas can be piped to refineries or power plants, or carried by tankers. They need relatively little processing before they are used, and they yield valuable chemical by-products.

Cons: All fossil fuels produce carbon dioxide, a 'greenhouse' gas that is responsible in the view of most scientists for global warming, and pollutants such as sulphur dioxide (responsible for 'acid rain'). But, above all, fossil fuels are 'finite' resources.

Generating energy 2

Chain reaction ❶

Albert Einstein's Special Theory of Relativity set the foundations for nuclear power in 1905, by showing how the **annihilation of mass** can release huge amounts of **energy**. In 1938 in Germany, Fritz Strassmann and Otto Hahn showed how this could be achieved through the **fission** (splitting) of uranium atoms by subatomic particles called neutrons.

A neutron hitting the nucleus of a uranium-235 atom splits the nucleus into lighter nuclei plus more neutrons, which can then go on in turn to make more U-235 atoms split. This creates a chain reaction. The fragments produced in this way weigh a little less than the original uranium nucleus, and the lost mass appears as energy – heat and radiation. As a result, the fission of 1 kg (2.2 lb) of uranium can yield as much energy as burning 2000 tonnes of coal or 8000 barrels of oil.

MIGHTY ATOMS Only the nuclei of uranium-235, or of man-made plutonium-239, take part in a chain reaction.

Nuclear waste ❷

Nuclear power produces large amounts of radioactive by-products. These wastes are classified as low, intermediate or high-level, depending on how intense and long-lasting their radioactivity is. **High-level waste** includes the fragments into which uranium nuclei split, such as strontium-90, caesium-137 and barium-140, and also plutonium and other heavy isotopes. **Reprocessing plants**, such as that at Sellafield in Cumbria, reprocess waste in order to produce new usable fuel.

Long-term plans include converting some wastes into a glassy, ceramic material that can be stored underground without – in theory – risk of contaminating water supplies.

WEIRD AND WONDERFUL ❸

In 1972, evidence was uncovered in West Africa of a natural nuclear reactor, 2 billion years old, in a uranium ore bed. Natural uranium would then have contained more uranium-235 than now – much has since 'burned up'.

Nuclear power around the world ❹

The chart on the left shows the top six countries in terms of the proportion of their electricity that was generated by nuclear plants in 2001. The chart on the right shows the top six in terms of nuclear power generation capacity in megawatts. The numbers in brackets show the actual number of nuclear plants – Lithuania's 78 per cent came from two reactors.

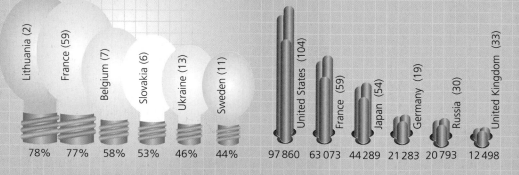

Lithuania (2)	France (59)	Belgium (7)	Slovakia (6)	Ukraine (13)	Sweden (11)	United States (104)	France (59)	Japan (54)	Germany (19)	Russia (30)	United Kingdom (33)
78%	77%	58%	53%	46%	44%	97 860	63 073	44 289	21 283	20 793	12 498

The main types of nuclear reactor ❺

The various designs use different methods and materials for moderating the nuclear reaction and extracting the heat. Figures are worldwide and apply to current operations (MW = megawatts)

PWR (pressurised water reactor)
◆ Most widely used type; 208 installed (capacity 198355 MW), plus 51 WWER (water-cooled, water-moderated energy reactor), a Soviet-era Russian pressurised water design (capacity 32834 MW).
◆ Moderator and coolant: water at high pressure (to prevent boiling).

BWR (boiling water reactor)
◆ Simplest and cheapest type; 90 installed (capacity 77878 MW).
◆ Moderator and coolant: water that is allowed to boil within reactor.

PHWR (pressurised heavy water reactor)
◆ Similar to PWR, but uses heavy water instead of ordinary water as moderator and coolant; 34 installed (capacity 16515 MW).
◆ Can use less-enriched or even natural (unenriched) uranium fuel.

GCR (gas-cooled reactor)
◆ One of the oldest designs, including Calder Hall (England; 1956); 18 installed (capacity 2930 MW), plus 14 AGR (advanced gas-cooled reactor; capacity 8380 MW).
◆ Moderator: graphite; coolant: carbon dioxide gas.
◆ GCR uses natural uranium fuel; AGR uses enriched uranium.

LWGR (light water graphite reactor)
◆ Soviet-era Russian design; 17 installed in former Soviet Union (capacity 12589 MW).
◆ Moderator: graphite; coolant: boiling water.
◆ The Chernobyl reactor that exploded in 1986 was of this type.

FBR (fast breeder reactor)
◆ Specially designed to create more plutonium fuel (for reactors or bombs) than uranium fuel 'burned'; 3 installed (capacity 1039 MW).
◆ No moderator; coolant usually molten sodium metal.

★ 548

No thanks!

Almost as soon as nuclear weapons were developed, movements like Britain's **Campaign for Nuclear Disarmament** began to protest about their potential for destroying all human life. Later, as the dangers of nuclear wastes became apparent, environmental movements campaigned against nuclear power, especially after the accidents at **Three Mile Island** (USA) in 1979 and **Chernobyl** (Ukraine) in 1986.

In 2001, **Germany** became the first major country to announce plans for the closure of all its nuclear power plants, promoting wind power in its place.

NUCLEAR PRODUCTION The four pressurised water reactors (PWRs) at Paluel in Normandy, France. The diagram below shows how they work.

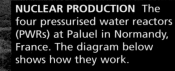

SHIELDING Thick concrete encloses the installation.

REACTOR CORE Contains fuel rods, control rods of cadmium or boron to absorb the neutrons, and the moderator.

PRESSURE VESSEL Surrounds the radioactive core of the nuclear reactor.

HEAT EXCHANGER Transfers heat from the coolant to water, generating steam.

STEAM TURBINE Steam drives the turbine, which drives an alternator, generating electric power.

A nuclear power station at work ❻

Uranium fuel is usually in the form of **rods** or **pellets**. A **moderator** – usually water, heavy water (containing deuterium rather than hydrogen) or graphite – is used to slow the fast-moving neutrons down so that they penetrate the uranium nuclei, causing the U-235 atom to split, rather than bouncing off them. **Control rods** are used to regulate power output or close the reactor down. A **coolant** – gas or liquid – circulates through the core to extract the heat and generated steam.

Generating energy 3

Tidal and wave power ①

The Earth's biggest energy reserves lie in the continuous movement of the oceans. There are two types of movement: the twice-daily rise and fall of the tides caused by the Moon's gravitational pull; and the much less predictable movement of waves, caused by wind.

Tides can be harnessed by a hydroelectric plant. Water flows into a dammed coastal lagoon at high tide, is held by the dam, and then flows out, passing through turbine generators. The **biggest working plant**, in the Rance estuary, France, opened in 1967.

Wave power is still largely experimental and used mostly in mid-ocean island sites such as Hawaii and the Azores, where waves are most constant. The total estimated wave potential, worldwide, is 2 to 3 million MW – equal to 2000 to 5000 nuclear power plants.

ISLAND ENERGY Waves strike Limpet 500, the world's first commercial-scale wave power station, on Islay, a Scottish Hebridean island.

Geothermal power ②

Colossal power is locked in the heat of the Earth's core, but is usable only in a few '**hot spots**' – mostly volcanic regions – where hot rocks or lava are found near the surface. In natural geysers, ground water boils to make steam, which can be collected to generate electricity. In other places, the hot rocks are dry; water can be pumped down to be heated, but this is not yet very widespread.

The **first geothermal plant**, at Lardarello, Italy, opened in 1904. The **biggest**, The Geysers in California, can generate up to 1900 MW. However, in many cases output has to be restricted so that the underground hot rocks do not cool too quickly. (Regeneration can take decades or even centuries.) There is also slight noise and air pollution.

The world's leading geothermal power producers (in order) are the USA, the Philippines, Italy, Mexico, Indonesia, Japan, New Zealand and Iceland.

★ 868

Biomass

Biomass consists of any **renewable organic material** – living or recently living plants and animals, their products and wastes, ranging from straw and paper to sugar and manure. Extracting energy from biomass is ecologically sound so long as the biomass is renewed. For example, by growing at least as many trees as are burned as fuel. It has been estimated that biomass methods could meet the world's entire energy needs.

Hydroelectric power ❸

Hydroelectric power – harnessing the energy of flowing or falling rivers – accounts for 18 per cent of the world's electricity, and 2.5 per cent of its total energy consumption. A hydroelectric plant can be installed at a **natural waterfall** – as at Niagara (built in 1893) – or by building a **dam** to create an artificial lake. The **world's biggest** plant is the Itaipu, on the Paraná river between Brazil and Paraguay; completed in 1982, its capacity is 12 600 MW. China's **Three Gorges** project on the Yangtze river, with a planned capacity of 18 200 MW, is due for completion in 2009.

HOOVER DAM Hydroelectric power plant on the Colorado River, USA.

Wind power ❹

Windmills have been used for well over 1000 years, mostly for pumping water or grinding grain. The **first wind electricity generator** was built in Denmark in the 1890s. In the 1920s a Frenchman, Georges Darreius, invented a wind turbine with three tall, curved blades looking like a giant egg-beater; it could catch the wind from any direction.

Most modern wind generators use huge **two- or three-bladed propellers**, up to 65 m (213 ft) across. They are computer-controlled to keep them facing the wind and to adjust the pitch (blade angle) according to the wind strength. They are usually grouped in '**wind farms**' on windy ridges or coasts, or offshore. The two biggest wind farms – with about 5000 turbines each – are in California. They are **pollution-free** and **cost-effective**.

FARMING THE WIND The Palm Springs wind generators in California, USA.

Tie-breaker ❺

Q: Which country is the world's leading wind generator?
A: Germany. Germany has more than 6000 MW of capacity – compared with 2500 MW in the USA – and more planned. Europe has nearly two-thirds of the world's total installed wind generation capacity.

Solar power ❻

All methods of energy generation except geothermal and nuclear power depend ultimately on the Sun. But solar power has only recently begun to be harnessed directly. Its most common use today is for heating domestic hot water in sunny areas, using roof-mounted solar panels. Electricity can be generated in two main ways:

◆ **Solar-thermal plants** use the Sun's rays, generally focused by mirrors, either directly to boil water to make steam, or to heat an intermediate liquid such as oil, which is then used to heat water. The latter system is used in the world's biggest solar power installation – a series of nine plants covering a total of 400 ha (1000 acres) in the Mojave Desert, California.

◆ **Solar-electric installations** use silicon photovoltaic (photoelectric) cells to turn sunlight into electricity. They are expensive but useful in isolated places – from space probes to motorway radio-telephones and lonely villages. They are also used in calculators and experimental cars.

HOME HEAT An array of solar panels (photovoltaic cells) on the roof of terraced houses in Heerhugowaaed in the Netherlands.

Ships

Core facts ❶

◆ A boat **floats** because the water it displaces – the water pushed aside by the hull – weighs the same as the boat itself, and pushes it up.
◆ A steel boat floats because it is hollow and can displace at least its own **weight of water**.
◆ A boat rides higher the less cargo it carries

and the denser the water. Cold, salty water is denser than warm and/or fresh water.
◆ The first craft were propelled by paddles or poles; later came oars and sails, followed by steam power, diesel, electric and gas-turbine engines, and nuclear power.

The ages of sail ❷

The first sailing craft were **'square-rigged'**, with a rectangular sail hung across the boat. This is

speedy downwind, but cannot sail into the wind. In the 5th century AD Arabs invented the triangular lateen sail – the first **'fore-and-aft'** rig, which can sail at an angle into the wind.

SQUARE RIG Ancient Egyptian boats used a simple, rectangular sail.

LATEEN RIG With a triangular sail on an angled spar the boat can sail at an angle to the wind.

CARAVEL Portuguese shipbuilders In the 15th century refined the Arab lateen rig.

GALLEON Combined square rig for downwind speed and lateen sails for manoeuvrability.

BARQUE This 'clipper', one of the fastest cargo sailing ships, had square and fore-and-aft sails.

KETCH A racing yacht is rigged fore-and-aft; it has a square-rigged spinnaker for downwind speed.

Underwater travel ❸

Submarines and **submersibles** have hulls strong enough to withstand 1 atmosphere (sea-level air pressure) for every 10m (33ft) of depth. Variable **buoyancy** (flotation) lets them sink or rise. Most submarines have water-filled ballast tanks that are 'blown' with compressed air to increase buoyancy. **Hydroplanes** (horizontal rudders to tilt the craft when it moves through the water) and/or adjustable propeller-like thrusters give 'dynamic' buoyancy.
Underwater power is usually electric or, in big submarines, nuclear.

NUCLEAR SUB Launched in 1954, USS *Nautilus* was the world's first nuclear-powered submarine.

GOING DOWN

9-12m (30-40ft) Normal limit for unaided diving; some divers can reach 30m (100ft).

30m (100ft) Limit for scuba (aqualung) diving using air.

700m (2300ft) Limit for diver using rigid diving suit.

900m (3000ft) Descent of Bathysphere manned by Beebe and Barton in 1934. Normal operating limit for a submarine.

6600m (21600ft) Limit for modern manned and unmanned deep-sea submersibles.

10911m (35797ft) World record dive, in steel bathyscaph *Trieste* (1960).

TIMESCALE

▶ **c.3200 BC** Egyptian plank-built boats.
▶ **2500-1200 BC** Minoan and Mycaenian narrow, oar-propelled war galleys and wide-beamed sailing cargo vessels.
▶ **c.700-500 BC** Fast Greek warships have two banks of oars (biremes) or three banks (triremes).
▶ **c.AD 400-500** Arab dhows have lateen sails.
▶ **8th-10th century** Viking longships with steering oar cross Atlantic.
▶ **Before 1200** Chinese invent rudder.
▶ **c.1200** Single-sailed cogs with high fore and stern 'castles' in northern Europe.
▶ **c.1450** First three-masted 'full-riggers' in Mediterranean, with square and lateen sails.
▶ **c.1540–1600** Highly manoeuvrable three and four-masted galleons.
▶ **17th century** Specialised warships and broad cargo ships (East Indiamen).

▶ **1787** First successful steam-powered vessel, a paddle-boat.
▶ **1800** First submarine with compressed air supply and hydroplanes.
▶ **1807** First commercially successful steamboat.
▶ **1836** Marine screw propeller patented.
▶ **1837** Brunel's *Great Western* – first ocean-going steamship, with side-paddles.
▶ **1845** Brunel's *Great Britain* – first propeller-driven ship to cross Atlantic.
▶ **c.1850** Three, four and five-masted 'clippers' are fastest-ever cargo ships under sail, at up to 21 knots (39 km/h).
▶ **1863** First engine-propelled submarine.
▶ **1864** First successful submarine attack on surface ship, in American Civil War.
▶ **1869** Opening of Suez Canal starts decline of clipper ships.
▶ **1897** First steam-turbine craft.
▶ **1898** First submarine with petrol engine (surface) and electric underwater motor.

▶ **1902** German *Preussen* – biggest-ever (steel) sailing ship, 132 m (433 ft) long.
▶ **1906** Britain launches *Dreadnought*, first modern battleship.
▶ **1908** First diesel-powered submarine.
▶ **1910** First diesel-powered seagoing surface ship.
▶ **1914-18; 1939-45** World Wars prove value of submarines (U-boats) in war.
▶ **1916** HMS *Argus*, first aircraft carrier. Austrian navy demonstrates hovercraft.
▶ **1937** First commercial hydrofoils.
▶ **1939-45** Second World War proves power of aircraft carriers over battleships.
▶ **1959** First full-sized hovercraft.
▶ **1975** *Nimitz*, first of US navy's 100000-tonne nuclear aircraft carriers.
▶ **c.1980** Japanese coastal vessels with rigid vertical wind foils save 50 per cent fuel.
▶ **1990** Hoverspeed pioneers operations of the world's first vehicle-carrying catamaran high-speed ferry, the Seacat.

FLOATING WORLD The ocean-going liner *The World* features 110 on-board luxury homes.

Hovercraft, hydrofoils and catamarans

Hovercraft ride on a cushion of air blown by powerful fans and trapped by flexible rubber 'skirts'. A **hydrofoil** has a foil – an underwater wing – that lifts the hull above the water. In a **catamaran**, or 'cat', the superstructure is held between two narrow hulls that cut easily through the water.

Records

Fastest clipper *Lightning* (1854): 436 nautical miles (807 km) in 24 hours; up to 21 knots (39 km/h).

Biggest liner *Voyager of the Seas* (1999): 311 m (1020 ft) long and 48 m (157 ft) wide; berths for 3800 passengers.

Fastest Atlantic crossing *Cat-Line V* (1998): 2 days 20 hours 9 minutes; 41.28 knots (76.37 km/h).

Deepest dive Jacques Piccard and Donald Walsh in bathyscaph *Trieste* (1960): 10911 m (35797 ft).

Largest ship The supertanker *Jahre Viking* (1980): 458.5 m (1504 ft) long.

483

Eureka!

The story goes that when **Archimedes** climbed into a bath and the water spilled over, he realised that any immersed object displaces its own volume of water. He went on to discover that the buoyancy (capacity to float) of an object in water equals and is balanced by the weight of the water it displaces – subsequently known as Archimedes' principle.

The numbers or star following the answers refer to information boxes on the right.

Motor cars and cycles

Core facts ❶

◆ There are almost 700 million **cars** on the world's roads today.
◆ **Mass production** has steadily reduced the real cost of motoring. The Model T Ford cost US$850 in 1908, but US$260 by 1925.
◆ In most advanced countries, there is one car for more or less every two people.

◆ Car ownership is growing by almost 5 per cent a year in the developing world, threatening a huge increase in road congestion and pollution.
◆ Despite research into alternative fuels and technologies, most cars use **petrol-fuelled** four-stroke **internal-combustion** engines.

Power cycle ❷

Most car engines run on the four-stroke engine. It gives a power stroke every two complete turns of the **crankshaft** (c).

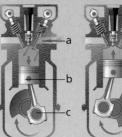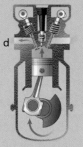

INDUCTION Inlet valve (a) open; **piston** (b) moves down. Petrol-air mixture is sucked into the cylinder.

COMPRESSION Piston moves up. Petrol-air mixture is compressed.

POWER Spark ignites fuel-air mixture. Burning gases force piston down.

EXHAUST Piston moves up. **Exhaust valve** (d) opens. Exhaust (burned) gases are blown out.

 ★ 397

Best sellers

In terms of the relative size of the car market at the time, the **Model T Ford** is the biggest seller, with over 15 million built between 1908 and 1927. About 22 million VW 'classic' **Beetles** have been made since 1936, but in 2000 this was overtaken by the **Toyota Corolla**: 29 million sold by January 2002.

Cars as stars ❸

◆ James Bond drove an **Aston Martin DB5** in *Goldfinger*, *Thunderball*, *On Her Majesty's Secret Service* and *Goldeneye*.
◆ In *Bullitt*, Steve McQueen drove a 1968 **Ford Mustang**.
◆ *The Italian Job* and *The Pink Panther* both featured the original version of the **Mini Cooper**.
◆ The cars in *The Love Bug* and its sequels were various **Volkswagen 'Beetle'** models; the first was a 1963 convertible.

BATMAN AND ROBIN The single-seat Batmobile appeared in the fourth Batman movie in 1997.

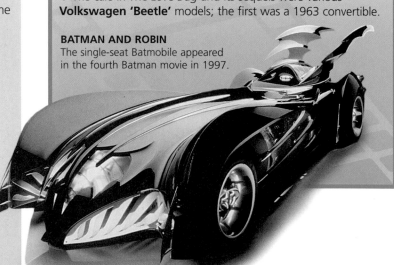

Future power ❹

◆ **Liquefied petroleum gas (LPG)** Less polluting but a finite fossil fuel; needs pressure tank.
◆ **Electric car (battery)** Needs frequent recharging; generating power for recharging may pollute.
◆ **Hybrid (electric motor plus petrol engine)** Complex but fuel-efficient, especially if motor reverses to recharge battery when braking; some emissions.
◆ **Electric car (solar cells)** Needs batteries for night-time use; non-polluting once built.
◆ **Hydrogen power (fuel cell, internal-combustion engine, or hybrid)** No direct emissions, but hydrogen generation uses electricity, which may pollute.
◆ **Biomass fuel (ethanol or biodiesel)** Renewable fuel; low greenhouse emissions.

Pedal power ❺

The bicycle developed from the 'Celerifère', a wooden scooter of the **1790s**. Baron von Drais added a seat and steering in **1818**, and in **1839** Macmillan added cranks and connecting rods. Then came the 'Vélocipède' with front-wheel pedals (**1861**), the high 'penny-farthing' (**1870**), and the 'Bicyclette' with same-sized wheels (**1879**). Derailleur gears moved the driving chain on sprockets (**1909**). Modern developments include the mountain bike (**1973**) and the use of carbon-fibre composites for racing bicycles in the **1980s**.

THREE WHEELS This tricycle from 1884 was Rover's first vehicle.

Motorcycles ❻

The first successful production two-cylinder motorcycle was launched in **1894**, followed in **1895** by a compact, high-speed four-stroke engine. In **1903** the Indian company (USA) launched the V-twin two-cylinder engine. Variable-speed gears were introduced in **1911** and the electric starter and drum brakes arrived in **1914**. The first small-wheeled motor-scooters (Vespa and Lambretta) appeared in **1947**. By **1959** Japanese makers had started to dominate the market and in **1968** Honda produced the first model with disc brakes. The **1980s** and **90s** saw the introduction of high-performance turbocharged and fuel-injected engines for some motorcycles.

MEAN MACHINE The V-Rod is the latest Harley-Davidson.

SOLAR POWER One of the cars taking part in the solar car race on the Suzuka circuit in Japan.

TIMESCALE ❼

▶ **1769-70** Nicolas Cugnot of France builds the first mechanically powered road vehicle, a steam-powered artillery tractor.
▶ **1860** Etienne Lenoir (France) builds first successful internal-combustion gas engine.
▶ **1883-6** Gottlieb Daimler and Karl Benz in Germany build first petrol engines.
▶ **1886** The first 'horseless carriage' – Daimler's four-wheeled petrol-driven vehicle.
▶ **1890** René Panhard and Emile Levassor (France) build the first purpose-designed car.
▶ **1892** German engineer Rudolf Diesel patents the compression-ignition engine.
▶ **1893** Wilhelm Maybach, an associate of Daimler, invents the modern carburettor.

▶ **1893** A steam-powered de Dion-Bouton wins first Paris-Rouen car race, averaging 18.7 km/h (11.6 mph).
▶ **1895** Panhard and Levassor build the first 'modern' car with a front engine driving the rear wheels. Michelin introduces pneumatic car tyres.
▶ **1900** First Daimler 'Mercedes', with a steel chassis.
▶ **1901-6** Oldsmobile is first mass-produced car.
▶ **1908** Henry Ford's 'Model T' launched.
▶ **1912** Cadillac is first with electric starter.
▶ **1913** Ford introduces moving assembly line to mass-produce Model Ts.
▶ **1920s** First diesel trucks and buses.
▶ **1922** Low-pressure 'balloon' tyres.

▶ **1924** Hydraulic brakes introduced.
▶ **1928** Syncromesh gearbox invented.
▶ **1935** Ferdinand Porsche designs the Volkswagen ('People's Car') for Hitler.
▶ **1939** Automatic transmission introduced.
▶ **1948** Michelin introduces radial-ply tyres, and Goodyear tubeless tyres.
▶ **1951** Disc brakes introduced.
▶ **1959** Front-wheel-drive 'Mini', with rubber suspension, launched in Britain.
▶ **1965** Ralph Nader's book *Unsafe at Any Speed* leads to stricter car safety standards.
▶ **Mid 1970s** First 'oil crisis' leads to emphasis on greater fuel efficiency in cars.
▶ **1980** Japan overtakes USA as car-maker.
▶ **1997** First highly fuel-efficient 'hybrid' petrol-electric cars produced in Japan.

Trains

Core facts ❶

◆ The train was the first true **mass transport** system, and a major social and economic force worldwide in the 19th century.
◆ There were over 9600 km (6000 miles) of track in Britain by 1850, and 14500 km (9000 miles) in the USA.

◆ At their peak in the early 20th century, railways carried most of the land freight and passenger traffic of the industrialised world.
◆ Railways remain important today for high-volume **commuter** transport, short-to-medium length **passenger** journeys and bulk **freight**.

The fundamentals ❷

The main principle of railways is the very **low friction** of steel wheels running on steel rails. Because of this, to transport a load by rail typically requires just one-tenth of the horsepower that would be needed to transport the same load by road.

There have been three main propulsion systems for railways:

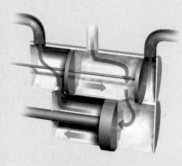  

STEAM POWER Steam, raised in a separate boiler, drives large pistons whose movement is transferred to the locomotive's wheels by cranks.

DIESEL POWER The ignition of diesel fuel, caused by pressure, drives the pistons in a large internal combustion engine.

ELECTRIC POWER Electric current picked up from a live rail or overhead wire drives a large electric motor that propels the locomotive.

Speed records ❸

Speed was a major selling point of trains from the very beginning. As railway technology advances, new records are still being set.

MALLARD The world record for steam locomotives remains unbroken to this day.

Fastest steam train Britain's *Mallard* set a record of 203 km/h (126 mph) in 1938.

Fastest train now in service French TGV (*Train à Grande Vitesse*) trains have run at up to 515 km/h (320 mph).

Fastest 'maglev' train Reached 552 km/h (343 mph) on a test track in Japan in 1999.

BULLET TRAIN Japan's Shinkansen, or Bullet trains, introduced in the 1960s, travel at up to 300 km/h (186 mph). **5**

Technology today

Railway technology is still advancing. Safety improvements, environmental concerns, road traffic congestion and competition with short-haul airlines all encourage new railway designs and technologies.

◆ **Automatic light railways** London's Docklands Light Railway uses small-size, driverless, computer-controlled trains on a rail network designed for minimum impact on the surrounding central city area.

◆ **High-speed trains** Streamlined trains with powerful electric engines and advanced control systems can reach high speeds on specially built tracks. Japan's Shinkansen (Bullet) and France's TGV are successful examples.

◆ **Magnetic levitation trains** 'Maglev' trains have been under development in Germany and Japan since the 1970s. These run at high speeds 1-10 cm (½-4 in) above a special track, supported by powerful electromagnets.

◆ **Monorail** Overhead monorails with suspended trains were invented in 1882, and have operated in Wuppertal, Germany, since 1902. Modern trains run on top of the rail, stabilised by guidewheels, in Sydney, Tokyo and other areas of high congestion.

◆ **Tilting trains** Trains that tilt (under computer control) for increased stability can take bends up to 40 per cent faster than conventional trains. Italy's Pendolino, Sweden's X-2000 and Spain's Alaris are tilting trains in current service.

'MAGLEV' Magnetic levitation trains are kept aloft by a powerful magnetic field.

Choo Choo **4**

'Chattanooga Choo Choo', one of the best-loved songs inspired by trains, was written for the 1941 movie *Sun Valley Serenade*. It was played by Glen Miller, with lyrics by Mack Gordon and music by Harry Warren.

★ 554

Going underground

The first tunnel for trains was opened in 1830, between Canterbury and Whitstable. The first entirely underground railway was London's Metropolitan Railway; its initial section, between Paddington and Farringdon, opened to the public on January 10, 1863. Electric trains were introduced on the London Underground in 1890.

TIMESCALE **6**

▶ **1712** Thomas Newcomen's first steam engine is built.

▶ **1760s** James Watt makes major improvements to the steam engine.

▶ **1801** Richard Trevithick's steam-powered carriage, *Puffing Devil*, is built; it runs on roads.

▶ **1804** Trevithick's first railway locomotive runs on a track at Pen Y Daren Ironworks, South Wales.

▶ **1825** George Stephenson's Stockton & Darlington Railway (the first public railway) opens, using his engine *Locomotion*.

▶ **1829** George Stephenson's *Rocket* wins the locomotive trials for the new Liverpool & Manchester Railway (opened 1830).

▶ **1830** The first US railway, the Baltimore & Ohio, opens, using the engine *Tom Thumb*.

▶ **1835** The first European railway opens, in Germany, using the Stephenson locomotive *Der Adler*.

▶ **1869** The US transcontinental railway is completed.

▶ **1879** Werner von Siemens demonstrates the first electric railway engine, in Germany.

▶ **1912** The first Diesel locomotive enters service, in Germany.

▶ **1916** The Trans-Siberian Railway, the world's longest line, opens after 25 years' work.

▶ **1964** Japan's Shinkansen (Bullet) trains enter service.

▶ **1991** British and French rail tunnels meet 23 km (14 miles) under the English Channel, forming the 50 km (31 mile) Channel Tunnel (rail services begin in 1994).

Flying machines

Core facts ❶

◆ The principles of **balloon** flight were developed by the French physcist and balloonist Jacques-Alexandre-César Charles (1746-1823).
◆ The principles of **aerodynamics** for heavier-than-air craft were first expounded by the English inventor Sir George Cayley (1773-1857) from 1809.
◆ Successful human flight also depended on the development of a light, powerful source of power, the **internal combustion engine** at the end of the 19th century.

Fundamentals of flight ❷

The wing's aerofoil shape creates **lift**. As the aircraft speeds along the runway, air flowing under the flatter lower surface moves a shorter distance and more slowly than the air passing over the curved upper surface. Air molecules under the lower surface bunch together, raising the air pressure. The **difference in pressure** above and below the wing pushes it upwards.

1 TAKE OFF In a modern aircraft wing, flaps at the trailing edge are lowered during take-off to increase the aerofoil effect.

2 LEVEL FLIGHT In level flight, the flaps are retracted to reduce drag. Smaller control surfaces (**ailerons**) in the outer wing are raised or lowered to bank the aircraft.

3 LANDING The flaps are lowered to maintain lift at low speed, and **spoilers** are raised from the upper wing surface to act as air brakes.

The jet engine

A practical **turbo-jet** for aircraft was developed by Frank Whittle in 1930-41 and Hans von Ohain in 1937-41. The world's first jet aircraft was the Heinkel He178 (1939).

Air is taken in at the front, compressed by a fan, mixed with fuel and ignited. The rapidly expanding gas that results escapes at high speed through the tail pipe (after turning the turbine that drives the fan). Its expansion simultaneously forces the entire engine – and thus the aircraft – forward through the air.

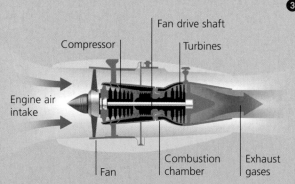

❸

Fan drive shaft

Compressor

Turbines

Engine air intake

Fan

Combustion chamber

Exhaust gases

TURBOFAN ENGINE Used in most modern airliners, this combines great power with fuel economy and a lower noise level.

SOLAR-POWERED
The unmanned *Helios Prototype* set a world altitude record for solar-powered aircraft on August 13, 2001, soaring to 29524m (96863ft).

The helicopter

◆ The helicopter's **rotor** is a spinning aerofoil, creating lift when it is rotated under power.
◆ Sophisticated **gearing** allows the rotor to be tilted in any direction, allowing the helicopter to move in the direction of **tilt**.
◆ **Torque**, the twisting force created by the spinning rotor, makes the helicopter's body rotate in the opposite direction; this is counteracted on single-rotor helicopters by a small, vertical tail rotor. Twin-rotor helicopters have rotors that turn in opposite directions, so their turning forces are cancelled out.

Balloons and airships

The principle of **balloon flight** is that an airtight bag filled with hot air, or a gas lighter than air (hydrogen or helium), will rise and can carry a load. **Airships** are powered balloons.
The Frenchmen **Joseph and Étienne Montgolfier** first demonstrated the balloon principle in 1783. Later in the same year, physicist **Jacques Charles** demonstrated the hydrogen-filled balloon. Another Frenchman, **Henri Giffard**, flew the first airship in 1852. The most successful operator of airships was Germany's **Zeppelin** company, from 1900 until the *Hindenburg* disaster of 1937.

NON-STOP *Breitling Orbiter III*, first to fly non-stop round the world.

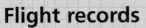

Flight records

Highest altitude for any aircraft 37 649.99 m (123 523.58 ft), on August 31, 1977, by Russian pilot Alexander Fedorov in a MIG-25.

Highest altitude for an aircraft in horizontal flight 25 929.03 m (85 068.99 ft), on July 28, 1976, by Americans Robert Helt and Larry Elliott in a Lockheed SR-71A Blackbird.

Highest altitude for a balloon 34 667.9 m (113 739.9 ft), on May 4, 1961, by US naval officers Malcolm Ross and Victor Prather in *Lee Lewis Memorial*.

Fastest speed for an aircraft over a straight course 3529.46 km/h (2193.17 mph), on July 28, 1976, by Americans Eldon Joersz and George Morgan in a Lockheed SR-71A Blackbird.

First aircraft to fly non-stop round the world without refuelling *Voyager*, December 14-23, 1986; designed by American Burt Rutan, piloted by Dick Rutan and Jeana Yeager.

TIMESCALE

▶ **1783** Jean-François Rozier and Laurent, marquis d'Arlandes, make the first successful human flight, in a balloon, over Paris.
▶ **1853** Sir George Cayley's model glider (1.5 m/5 ft long) is the first true aeroplane to fly.
▶ **1903** Orville Wright makes the first controlled, sustained flight by a powered aeroplane at Kitty Hawk, North Carolina, USA.
▶ **1909** Louis Blériot crosses the English Channel in a monoplane.
▶ **1919** John Alcock and Arthur Whitten Brown make the first non-stop transatlantic flight, in a Vickers Vimy biplane. Britain's R34 makes the first transatlantic airship crossing.
▶ **1927** Charles Lindbergh makes the first solo non-stop transatlantic flight, from New York to Paris.

JUMBO CARRIER
The improved Boeing 747-400 'jumbo jet' came into service in 1989.

▶ **1928** Australians Charles Kingsford-Smith and Charles Ulm make the first flight across the Pacific, from San Francisco to Brisbane.
▶ **1936** The Douglas DC3 ('Dakota') airliner enters service.

▶ **1937** The airship *Hindenburg* catches fire at Lakenhurst, New Jersey, effectively ending the use of airships for passenger transport.
▶ **1939** The first jet aircraft, the Heinkel He178, flies in Germany. Igor Sikorsky designs the first successful helicopter.
▶ **1949** The first jet airliner, the De Havilland Comet, enters service.
▶ **1966** The Hawker Siddeley Harrier, the first successful vertical take-off and landing aircraft, enters service in Britain.
▶ **1968** The Tupolev Tu-144, the first supersonic airliner, enters service in the USSR.
▶ **1970** The Boeing 747 'jumbo jet' enters service.
▶ **1976** The Anglo-French Concorde, the most successful supersonic airliner, enters service.
▶ **1979** A pedal-powered aircraft, *Gossamer Albatross*, flown by cyclist Bryan Allen, successfully crosses the English Channel.

▶ **1981** *Solar Challenger* makes the first long-distance flight by a solar-powered aircraft.
▶ **1999** *Breitling Orbiter III*, piloted by Bertrand Piccard and Brian Jones, is the first balloon to circle the world non-stop.

⭐ **414**

Gliders

Gliders are unpowered, taking their lift purely from airspeed and naturally rising hot air currents (**thermals**) – they evolved from the kite. George Cayley of Britain and Otto Lilienthal of Germany were pioneers in the 19th century.

Into space

Core facts ❶

◆ Exploration of **near-Earth space** began in 1957 and continues today. Exploration beyond the Solar System remains a distant prospect.
◆ The American physicist **Robert Goddard** and the German engineer **Wernher von Braun** separately pioneered practical rocketry from the late 1920s onwards.

◆ Space programmes were initially the preserve of the **USA** and the **USSR**, driven by Cold War rivalry. Over a dozen countries are now involved in space research.
◆ The pre-eminent space exploration body today is the USA's **National Aeronautics and Space Administration (NASA)**.

Principles of space travel ❷

The principles of space travel were laid out by Russian physicist **Konstantin Tsiolkovsky** in a book published in 1903. He proved that rocket propulsion would work in space, calculated the escape velocity for a rocket to overcome Earth's gravity (40 000 km/h or 25 000 mph), and proposed liquid fuel and **multistage rockets**, with initial stages jettisoned after use.

Once beyond significant gravitational pull from Earth, a spacecraft travels by momentum, steered by its own small rocket engines. Spacecraft require robust communications and life-support systems, and extensive heat-shielding to prevent them burning up on re-entry to the Earth's atmosphere.

The rocket can escape the Moon's much smaller gravitational pull at 8640 km/h (5370 mph)

The rocket escapes Earth's gravitational field at a speed of 40 000 km/h (25 000 mph).

BREAKING FREE To gain speed, rockets are launched near the Equator and in the direction of the Earth's spin, which helps to 'throw' them into space.

The space shuttle ❸

The space shuttle is the first **reusable spacecraft**. It carries payloads of up to 29 tonnes into orbit, with a crew of two to eight people. Five shuttles have been used for space travel: *Columbia*, *Challenger* (exploded 1986), *Discovery*, *Atlantis* and *Endeavour*.

For take-off, the shuttle has two **booster** rockets and an external fuel tank, all jettisoned before the craft reaches orbit. It re-enters the Earth's atmosphere belly-first, protected by a **ceramic heat shield**, and glides to an aircraft-style landing.

Circling the Earth ❹

An **artificial satellite** is an object placed in orbit around the Earth or another body in space. Satellites are launched by rockets or by the **space shuttle**. Above about 320 km (200 miles), satellites remain in orbit without propulsion. At 36 000 km (22 300 miles) high, an orbit takes 24 hours: the satellite is **geosynchronous** – it synchronises with the Earth's rotation. More than 1000 artificial Earth satellites now exist. Their main functions are navigation, weather monitoring, communications, observation and scientific research.

SATELLITE STATION The Russian-built *Zarya* control module of the International Space Station, launched in 1998.

How rockets work

Rockets are the only means of propulsion in the **vacuum** of space. Huge pressure is created by burning vast quantities of fuel, either **solid** (aluminium oxide) or **liquid** (hydrogen), with oxygen. At the lower end of the **combustion chamber**, gases escape through a nozzle, so that the pressure at the top is unequalised; this forces the entire rocket upward. A multistage rocket is effectively several rockets that operate in succession.

Area occupied by astronauts shown in yellow

ON THE PAD The progress of rocket science, from the German V-2 rocket (a, 1945), to the US Atlas-Mercury (b, 1962), US Titan-Gemini (c, 1964), Soviet A2-Soyuz (d, 1967), US Saturn V-Apollo (e, 1967) and the space shuttle (f, 1981).

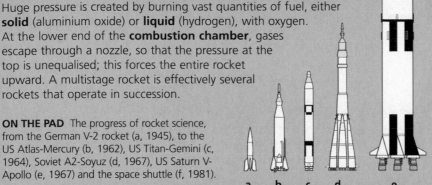

a b c d e f

Space probes

Uncrewed space missions ('probes') provide data without risking human lives. They can continue transmitting information as they travel towards outer space, or can land on other planets. Some important missions include:

Venera 7
Soviet probe, launched August 17, 1970. Landed on **Venus** December 15, 1970, and sent weak signals for 23 minutes. First manmade object to land on another planet.

Voyager 1 & 2
US probes, launched September 5 and August 20, 1977. Sent images of **Jupiter**, **Saturn**, **Uranus** and **Neptune** (1979–89). In 1998 *Voyager 1* became the most distant manmade object in space.

Galileo
US probe, launched October 1989. The lander was lost on **Jupiter** on December 7, 1995, after sending data for a hour during descent. The orbiter still sends images of the planet and its moons.

Pathfinder
US probe, launched December 1996. Landed on **Mars** July 4, 1997. Data was transmitted by the probe's *Sojourner* surface explorer module until contact was lost September 27, 1997.

Cassini-Huygens
US/European probe, launched October 1997. The *Cassini* orbiter will reach **Saturn** orbit July 2004. The European Space Agency's *Huygens* module will probe the atmosphere of the planet's moon Titan from November 2004.

303

Down to Earth

Space exploration has had many spin-offs, including quartz watches, the artificial heart, pressurised ball-point pens, cordless electric drills and charge-coupled devices (CCDs), used in digital photography.

MARATHON MAN 'Space blankets' are made from the material used for external insulation on space missions.

Weapons 1

Core facts ❶

◆ Warfare has been one of the driving forces behind **technological change** from the ancient world onwards.

◆ **Early advances** in military technology depended mainly on developments in metallurgy and metalworking.

◆ The greatest change came with the development of **firearms**, from the 15th to the 16th centuries.

◆ Early weapons mostly required soldiers to **fight face to face**. Now most weapons work **remotely**.

The firing range ❷

The earliest projectile weapons were the **bow** and the **sling**, both in use from prehistory. The wooden bow evolved into the more powerful composite bow (strengthened with horn and sinew), the **crossbow** and the **longbow** by the later Middle Ages. Firearms, first mentioned as 'hand gonnes' in about 1400, were unreliable until the **flintlock** came into widespread use from the 1660s. It remained the standard infantry weapon for 150 years. The first **artillery** consisted of devices such as catapults, which threw stones or metal arrows (bolts). Siege engines were used to batter the walls of enemy fortresses. The first **cannon** came in the 14th century. Heavy siege cannon were joined by lighter **field guns** on wheeled carriages, for use on the battlefield.

BACKGROUND IMAGE Siege warfare in a relief from Nineveh, showing an Assyrian attack on Israelite Lachish, 587-586 BC.

GROWLING TIGER An Indian bronze mortar from the late 18th century, shaped like a tiger.

SPIKED DEATH A lethal flail.

Hand-to-hand combat

The oldest hand weapons are the spear, club, mace and axe, which all developed from hunting equipment. The **sword** became effective as a weapon about 1200 BC, with the development of iron working. Most swords were short until the long, heavy *spatha* was adopted in late Roman times. The medieval sword was crafted with increasing skill: it was hammer or pattern-welded and repeatedly fired to convert iron to steel. The sword remained the

❸

pre-eminent hand weapon, despite the proliferation of pole-arms, such as **bills** and **halberds**, used by infantry against mounted knights. Weapons like the **mace** and **flail**, with spiked heads, were for smashing an opponent's armour.

The advent of firearms made armour obsolete, so swords became lighter and better adapted to the defensive parry – ultimately the long **rapier** of the early 17th century. The slashing cavalry **sabre** evolved from Asian and Middle Eastern swords like the scimitar, which were curved to increase their cutting power. Repeating firearms like the revolver and machine gun finally relegated the sword to ceremonial use in the late 19th century.

KNIGHT KILLER The bill developed from the agricultural bill hook. Italian craftsmen made this one in the 16th century for Henry VIII.

TIMESCALE ❹

▶ *c.*30000 BC Spears, clubs and the bow and arrow are all in use.
▶ *c.*4000 BC Egyptian soldiers make the first known use of shields.
▶ *c.*2500 BC Sumerian soldiers wear the first known armour.
▶ *c.*1600 BC The war horse is first used, in Persia.
▶ *c.*1500 BC The Greeks invent bronze body armour.
▶ *c.*500 BC The composite bow is invented in Persia, and the crossbow in China.
▶ *c.*400 BC Mechanical bolt-throwing artillery is developed in Greece.
▶ 1st century BC Rome's legions are equipped with mobile artillery and standardised armour and weapons.
▶ 668 'Greek fire', an incendiary weapon, is invented by the Byzantines.
▶ *c.*950 The first motte-and-bailey castles are built in medieval Europe.
▶ 1241 The Mongols make the first use of gunpowder in European warfare, at Sajo in Hungary.
▶ Late 13th century The longbow is developed in the British Isles.
▶ *c.*1460 The matchlock mechanism is developed in Europe, starting the rise of firearms as the dominant battlefield weapon.
▶ 1647 The bayonet is invented in France, making infantrymen without firearms redundant on the battlefield.
▶ Late 17th century Low, thick-walled fortresses designed to withstand siege artillery spread throughout Europe and the European colonies.
▶ 1792 Rocket projectiles are first used, by Mysore (India) forces against the British.
▶ *c.*1850 Breech-loading rifles begin to replace muzzle-loading muskets.

The cavalry ❺

For most of history, the expense of the **war horse** made cavalry service the preserve of the wealthy. Mounted soldiers were not effective in the ancient world – mainly because their numbers were small. But the **armoured knight** dominated medieval warfare until the rise of firearms. The horseman with breastplate and sabre enjoyed a brief revival in the Napoleonic wars of the early 19th century. The **last cavalrymen** saw brave but futile action in Poland at the start of the Second World War in 1939.

Shields and armour ❻

The **shield** played an important military role until the end of the Middle Ages. **Helmets** and **body armour** were rare in antiquity, but were used widely from the Roman period to the end of the Middle Ages.

Scale armour (small overlapping plates of metal attached to a leather garment) was the earliest type, made as early as 500 BC. The oldest **chain mail** (flexible armour of chain-like links) dates from about 450 BC, from Kiev – but it was common only from the late Roman period until about the 14th century. **Plate armour**, made from plates of iron or steel, offered the best protection. By the 15th century the most expensive suits were very elaborate and mechanically complex.

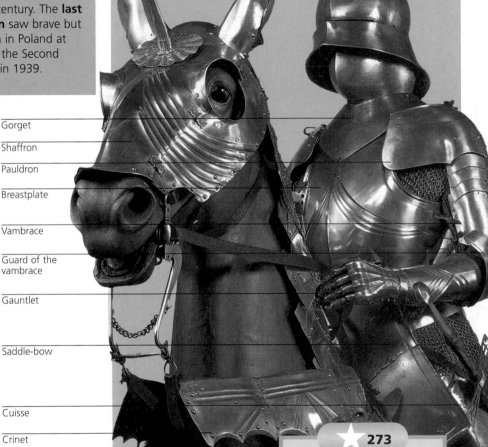

Gorget
Shaffron
Pauldron
Breastplate
Vambrace
Guard of the vambrace
Gauntlet
Saddle-bow
Cuisse
Crinet
Peytral

FULL ARMOUR By the late 15th century, plate armour had reached a peak of sophistication, with a special name for each of the many different plates. This suit of armour for a knight and his horse was made in Germany.

⭐ 273

Musketry
The first shoulder-fired firearm was the hackbut or **harquebus**, invented in Spain in the mid 15th century. It was superseded by the larger **musket**, also from Spain. Early muskets fired 57 g (2 oz) balls about 160 m (525 ft), but with poor accuracy.

Weapons 2

Rapid fire ❶

The American inventor Hiram Maxim developed the **first fully automatic machine gun** in 1884. The Maxim gun used the energy of each recoil to load the next cartridge. The Thomson hand-held **submachine gun** (the 'Tommy gun') was invented during the First World War. The Second World War saw the distinction between two-man, bipod-mounted light machine guns and the tripod-mounted heavy machine gun. Infantry rifles and machine guns converged after the Second World War with the development of **assault rifles**, such as the AK-47, capable of both semi-automatic (firing and reloading automatically once on each trigger-pull) and fully automatic action (continuous fire).

POPULAR CHOICE
The AK-47 can fire 600 rounds a minute.

Rocket propelled ❷

From Germany's V-2 flying bomb, first used against Britain in 1944, rocket-powered missiles have become one of the most potent modern weapons systems. The table lists typical examples of three generations of ballistic missiles:

Missile	V-2	SS-6	Trident 2
Country	Germany	USSR	USA
Date introduced	1944	1957	1990
Type	Single stage	Two stage	Three stage
Fuel type	Liquid oxygen	Liquid oxygen	Polythylene glycol (a solid fuel)
Maximum range	320 km (200 miles)	13 000 km (c.8000 miles)	7360 km (4600 miles)
Payload	725 kg (1600 lb) high explosive	3.5 megaton thermonuclear	Thermonuclear MIRV (various configurations)
No of warheads	One	One	Multiple
Notes	Ground launched; the first true ballistic missile	Ground launched; the first multistage intercontinental ballistic missile (ICBM)	Launched from a submerged submarine; currently deployed by the USA and UK

Armoured vehicles ❸

The tank (a name given to disguise the project's true purpose), or armoured fighting vehicle, was developed in Britain during the First World War, in an attempt to break the trench warfare deadlock by technology. The **first tanks** saw service at the Somme (1916) and at Cambrai (1917). In the Second World War, tanks spearheaded the German Blitzkrieg and played a major role in all European armies; in 1944, the year of highest production, 51 128 Allied and 19 002 German tanks entered service. Tanks remain the most important ground-assault weapons in modern warfare, with computerised navigation, targeting and gun-stabilisation systems, powerful engines and protection against shells, rocket projectiles, gas and biological weapons – and a price to match.

British Mark IV, used in the First World War

Soviet T-34, used in the Second World War

American M1A1 Abrams used in the Gulf War

ATOMIC TEST The first of 23 US atmospheric tests took place on Bikini Atoll in the Pacific, in July 1946.

Nuclear weapons ❹

The first **atomic** bomb, containing 60 kg (130 lb) of fissile uranium, was dropped on Hiroshima, Japan, in 1945. It exploded with the force of 15 000 tons of TNT, destroying 67 per cent of the city and killing 66 000 people instantly.

Thermonuclear bombs, known as hydrogen bombs because they work by fusion of the hydrogen isotopes deuterium or tritium, have produced test explosions equivalent to 50 megatons (50 million tons) of TNT. They were first tested in 1952.

The **neutron** bomb is a small thermonuclear weapon that produces a limited explosion but a huge wave of lethal (though short-lived) neutron and gamma radiation.

★ **278**

Smart weapons

Smart weapons are munitions – usually missiles – with computer guidance systems that enable them to pinpoint a target. A **laser-guided missile**, which directs itself toward a target picked out by a laser aimed from a circling aircraft, is one example. The **cruise missile**, which tracks ground contours to follow a pre-set route to its target, is another.

War in the air ❺

Air warfare – strategic bombing, ground-attack aircraft and fighter combat – developed during the First World War. On August 20, 1914, the first Zeppelin airship raids took place on London. A British BE2 dropped the first bomb from an aircraft on August 26.

The Second World War saw dive-bomber support of the German Blitzkrieg, 'carpet bombing' of enemy positions and carrier-based aircraft triumphing over battleships.

In recent wars, the combination of attack helicopters, cruise missiles and high-altitude heavy bombing has proven decisive, as in Afghanistan in 2001-2.

TIMESCALE ❽

▶ **1862** The hand-cranked Gatling gun – the first effective machine gun – was patented.
▶ **1880** Breech-loading artillery firing self-contained shells is generally adopted.
▶ **1897** France introduces the 75 mm quick-firing field gun, the first with on-carriage recoil and a shield for the gunners.
▶ **1914-18** Tanks, gas, flamethrowers and air warfare are first seen during the First World War.
▶ **1940** The first air-defence rockets are developed, in Britain.
▶ **1942** The bazooka infantry anti-tank weapon is used by the US army.
▶ **1944** Germany develops the V-1 pilotless flying bomb (known as the 'Buzz Bomb' by its British targets).
▶ **1950** Military helicopters are first used, during the Korean War.
▶ **1957** The USSR and USA develop intercontinental ballistic missiles.
▶ **1991** The Gulf War sees the first widespread use of 'smart' munitions.

STEALTH FIGHTER ❻
The USAF's Lockheed F-117A Nighthawk fighter-bomber, operational from 1983, is designed to be invisible to radar.

War at sea ❼

The earliest sea warfare transferred land weapons and techniques (such as bows and arrows and boarding parties) to ships. Specifically naval innovations included the ram bow of the Greek galleys and the raised 'castles' for archers on medieval warships. Guns were carried by the galleons of the 16th century, leading to the line-of-battle ships of the 18th.

Armour plating, steam propulsion and large guns in turrets revolutionised naval warfare in the 19th century; Dreadnought battleships (from 1906), submarines (from about 1905) and aircraft carriers (from 1918) in the 20th. Submarine-launched nuclear cruise missiles and ballistic missiles are among the most potent of modern weapons.

Other weapons ❾

Gas was first used in the First World War, starting with the German use of gas shells against Russian lines at Bolimov in 1914. Countermeasures such as gas masks limited the use of gas in the Second World War. Today, many countries have **chemical weapons**, but their use is limited by international convention. The USA is developing '**non-lethal**' weapons – such as lasers that cause temporary blindness and sonic weapons that cause severe pain by inducing vibrations in internal organs.

CHEMICAL DANGER A US marine equipped for chemical warfare during the Gulf War in 1991.

Electromagnetic spectrum

Core facts ❶

◆ Electromagnetic radiation is best understood as **waves of energy**, produced by oscillating electric and magnetic fields.
◆ Physicists divide this energy into a **spectrum** of named bands, according to frequency and wavelength. In this spectrum, our senses detect only heat (**infrared**) and visible **light**.

◆ In a vacuum, all electromagnetic waves travel at a speed of roughly 300 000 km (186 000 miles) per second – the **speed of light**.
◆ Electromagnetic waves can also travel through **matter**, at a speed reduced by up to a half.
◆ **James Clerk Maxwell** laid out the basic theories of electromagnetism in the 1860s.

RADIO WAVES
Up to 3000 MHz
Used in radio and television broadcasting, telecommunications networks, cordless telephones, wireless computer networks, radio telescopes.
 The usable spectrum of radio frequencies is divided into 'bands' and regulated nationally and internationally to avoid interference between broadcasters.

MICROWAVES ❷
3000 MHz to 300 GHz
Used in telecommunications networks, microwave ovens, mobile telephones, radar, microwave telescopes.
 Absorbed by water molecules in food to 'cook' it in a microwave oven, but not absorbed by plastics or ceramics.

INFRARED ❸
300 GHz to 400 THz
Used in electric heaters, cookers, toasters, remote controls for televisions and VCRs, aerial surveying, night-vision equipment, thermal-imaging cameras, infrared astronomy.
 Produced by warm or hot objects, and generally identifiable as 'heat'; detected by the eyes of some animals; causes warming when absorbed.

VISIBLE LIGHT ❹
400 to 750 THz
Used in photography, cinematography, TV, power generation (by photocell).
 The range of radiation detected by human eyes; powers photosynthesis in plants.
 Splits into the colours of the rainbow: red, orange, yellow, green, blue, indigo and violet.

Newton's light ❺

Until the 17th century, white light was thought to be a simple, single colour. From about 1666, however, the young **Isaac Newton** conducted a series of experiments in optics, noting that a thin beam of white light is split into an elongated **spectrum** of the rainbow colours by a glass **prism**.
 He concluded that white light is made up of rays of all colours, which are **refracted** at slightly different angles by the prism. His conclusions were published as *Opticks* in 1703.

Frequency and wavelength ❻

◆ **Frequency** is the number of times an electromagnetic wave reaches its peak (vibrates or oscillates) per second. It is measured in **hertz** (Hz).
◆ **Wavelength** is the distance an electromagnetic wave travels in the course of one oscillation. It is measured in **metres** (m).
◆ Higher frequencies mean shorter wavelengths and shorter wavelengths mean greater energy: **wavelength = the speed of light ÷ frequency**.

Colour ❼

Colours are light of different wavelengths, ranging from red with the longest wavelength to violet with the shortest. Sensors in the **retina** (the inner surface of the eye) actually respond only to wavelengths corresponding to red, green and blue (the **primary** colours), but the brain interprets combinations of these as the thousands of colours we perceive.

There are two kinds of primary colours: **additive** and **subtractive**. Television screens, for example, produce colours by mixing glowing phosphor dots in the three **additive primary colours**: red, green and blue.

Colour printing produces colours by mixing the **subtractive primary colours**: cyan (blue-green), magenta (blue-red) and yellow – often referred to simply as blue, red and yellow. Combinations of these inks absorb most light wavelengths, reflecting only those of the required colour.

Photons and other quanta

Photons are particles, or **quanta**, of light (quanta are minute discrete increments into which electromagnetic radiation is subdivided).

Photons are one of the **elementary particles** known in quantum physics as gauge bosons.

Their existence was predicted by **Max Planck** and **Albert Einstein** in 1900-05 to account for the particle-like behaviour of light in certain conditions.

Waves and walls ❿

Most types of electromagnetic radiation are **potentially dangerous** to humans, including microwaves and infrared (heat) as well as ultraviolet, X-rays and gamma rays.

The ability of any material – including human skin – to resist penetration is determined by the energy and the wavelength of the radiation, as well as by the density of the material.

Short-wavelength, high-energy radiation, such as X-rays or gamma rays, can penetrate all but the densest shielding and cause damage to living cells. Extremely thick concrete is needed as a shield from sources of **gamma rays**, such as nuclear reactors.

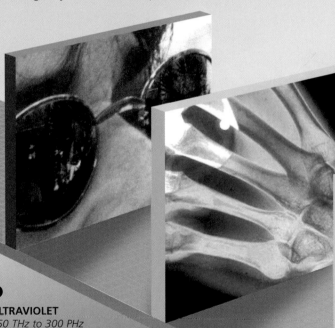

❽

ULTRAVIOLET
750 THz to 300 PHz
Used in some night-vision equipment, hospitals (for sterilisation of equipment), optical brighteners in washing powder.

Causes suntan and sunburn in humans, and some cancers; can be detected by insect eyes; intense ultraviolet rays from the Sun are absorbed by the ozone layer in the Earth's atmosphere.

X-RAYS *300 PHz to 30 EHz*
Used in X-ray equipment (for observing dense internal body structures, like bones), checking joints and welds in metal, destroying cancer cells.

X-rays pass through many solids, including human soft tissue; can cause genetic mutation and cancers; can damage some electronic equipment.

GAMMA RAYS *More than 30 EHz*
Used in checking dense structures in engineering, destroying spancer cells.

Produced by nuclear reactions, including both nuclear explosion and radioactive decay; can pass through all but the most dense solids; causes genetic mutation, cell damage and 'radiation sickness'.

Observing the Universe

Core facts ❶

◆ Visual observation is limited by the eye's **inability to focus** on astronomically distant or microscopically close objects. Lenses in telescopes and microscopes make this possible.
◆ Lenses work by **refraction** – the fact that light bends as it passes from one medium (air, for example) to another (the glass of a lens).

◆ **Mirrors** aid visual observation (for example, in reflecting telescopes) because reflection can be used to focus light into an image.
◆ **Electron microscopes** allow us to observe tiny objects invisible to human eyes, while **radio telescopes** provide visual images of non-visual phenomena such as radio wave emissions.

Spectacles ❷

◆ Spectacles use **refraction** to correct the inability of many eyes to focus light correctly on the surface of the retina.
◆ **Convex** lenses were developed in the late 13th century in Italy and China.
◆ **Concave** lenses were first made in the mid 15th century.
◆ **Ophthalmology**, the scientific matching of lenses to an individual's vision, was developed in the late 17th century.

CONVEX LENS
This corrects hypermetropia (longsightedness) by shifting the focal point forward from behind the retina.

CONCAVE LENS
This corrects myopia (short-sightedness) by shifting the focal point back from in front of the retina.

Today's microscopes ❸

Advances in physics have led to a number of alternatives to the optical microscope. These magnify 1 000 000 times and more, and have even allowed imaging at the level of individual atoms.

Electron microscopes were developed from the 1930s onwards. The **transmission electron microscope** (TEM), invented in 1932, passes a beam of electrons through a specimen sliced extremely thinly, creating an image on a fluorescent screen or photographic plate. The **scanning electron microscope** (SEM), invented in 1935 but used only from about 1965, moves a beam of electrons over the surface of a specimen held in a vacuum chamber.

Further developments include the **scanning tunnelling microscope** (STM) and the **atomic force microscope** (AFM), invented in the 1980s. These use a probe with a tip as small as a single atom to scan the surface of the specimen.

Another development is the **acoustic microscope**, which uses ultrasound to scan the specimen.

SURFACE SHOT The surface of a polymer film, imaged using an atomic force microscope.

Optical microscopes ❹

Compound optical microscopes make use of the magnifying effect of the convex lens. They differ from a magnifying glass by using **two or more lenses**: one lens, the objective, magnifies the object; an eyepiece lens, the ocular, magnifies the image of the first lens. Magnifications typically range up to 1500 times. Specimens are normally mounted on a glass slide and lit from behind.

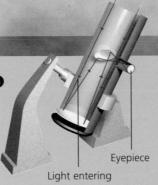

Optical telescopes ❺

◆ **Refracting telescopes** use a large lens, the objective, to focus the image into a magnifying ocular lens.

◆ **Reflecting telescopes** use a concave mirror to focus the image, which is directed into an eyepiece. The mirror eliminates chromatic distortion (colour distortion through the prism effect of a lens).

◆ **Binoculars** consist of two small telescopes mounted in a single frame, providing a natural, stereoscopic image – an image with depth of field – of distant objects.

Eyepiece

Light entering

REFLECTING Light enters the tube and is reflected by a series of mirrors into the eyepiece.

⭐ 128

Hubble vision

The **Hubble Space Telescope** (HST) is a powerful reflecting telescope system orbiting 966 km (600 miles) above the Earth's surface. Lying beyond the distorting effect of the atmosphere, it provides a resolution up to ten times greater than that of any telescope on Earth. It was launched into orbit in 1990; defects in its operation were corrected in a space shuttle mission in 1993.

BACKGROUND IMAGES: MACRO AND MICRO This page: A photo of the Cone Nebula, 2500 light years away, taken using the Hubble Space Telescope. Opposite: An image of the scales on a butterfly's wings made using a scanning electron microscope.

Galileo ❼

The Italian Galileo Galilei (1564-1642) was the first person to use the newly invented refracting telescope to **study the stars**. With it, he discovered the irregularity of the Moon's surface; identified Jupiter's moons; confirmed that the Milky Way was made up of stars; and, through observing sunspots, proved the theory of Copernicus that the Earth revolved around the Sun. An Inquisition court found him guilty of heresy for the last discovery, because it ran counter to the Church's teaching that the Earth was the centre of the Universe.

Today's telescopes ❻

The world's largest refracting telescope is at **Yerkes Observatory** in Wisconsin, USA. Its objective lens has a diameter of 102 cm (40 in).

The largest single reflecting telescope is at **Zelenchukskaya** in Russia. Its mirror has a diameter of 6 m (236 in). The **Keck** telescopes at Mauna Kea, Hawaii, use segmented mirrors, with a diameter of 10 m (33 ft) each.

Since the discovery of cosmic radio emissions in the 1930s, **radio telescopes** have opened a new field of astronomy. They use a large metal bowl or dish as an antenna to focus radio waves. The Lovell Telescope at **Jodrell Bank** in Britain is a radio telescope, with a 76 m (250 ft) dish.

The largest single radio telescope dish, at **Arecibo**, Puerto Rico, has a diameter of 305 m (1000 ft). But best results are obtained from multiple-dish systems like the **Very Large Array** (VLA) at **Socorro**, New Mexico, USA.

BIG DISH The Arecibo telescope's antenna lies in a volcanic crater, lined with steel reflectors (below).

TIMESCALE ❽

▶ **1289** The earliest written record of spectacles occurs in a document of Sandro di Popozo of Italy.

▶ **1590s** Dutch spectacle makers Hans and Zacharias Janssen make the first compound optical microscope.

▶ **1608** Dutch lensmaker Hans Lippershey invents the refracting telescope.

▶ **c.1670** British physicist Isaac Newton makes the first reflecting telescope.

▶ **1785** American scientist and statesman Benjamin Franklin invents bifocals.

▶ **1887** Swiss doctor Adolf Eugen Frick makes the first (hard) contact lenses.

▶ **1932** German scientists Ernst Russka and Max Knoll invent the transmission electron microscope.

▶ **1935** Max Knoll invents the scanning electron microscope, though it is not commercially developed until the 1960s.

▶ **1937** US astronomer Grote Reber builds the first radio telescope.

▶ **1970s** The acoustic microscope is developed in the USA.

▶ **1981** IBM researchers Gerd Binning from Germany and Heinrich Rohrer from Switzerland invent the scanning tunnelling microscope, capable of resolving individual atoms.

Navigation and detection

Core facts ①

◆ During the 20th century, highly sophisticated ways of locating objects were developed. Most of these developments were for military use, but civilian applications quickly followed.

◆ **Sonar** and **radar** were developed to detect unseen enemy equipment from reflected sound or radio waves – sonar for submarines, radar for surface ships or aircraft in flight.

◆ Both **radar** and **radio navigation** make use of the fact that a loop aerial transmits radio waves in a narrow beam, rather than through 360 degrees.

◆ **Sonar** uses high-frequency sound waves because these travel very quickly and efficiently in water – radio waves do not travel through water at all.

What is radar? ②

Radar is a system for locating objects and navigation by transmitting **high-frequency radio waves** and detecting their reflection (echo).

Pulse radar, the most common type, transmits short bursts of radio waves and calculates the distance to a target from the time it takes for the echo to return. A directional antenna means that the bearing of the target can also be calculated.

The **development of radar** occurred independently in Britain, France, Germany, the USA and Japan in the 1930s.

Radar is now also used in ICBM early-warning systems and missile guidance, air traffic control, shipping, astronomy, meteorology, archaeology, satellite tracking and speed detection by traffic police.

BRITISH PIONEER Robert Watson-Watt (above) oversaw the development of Britain's first effective anti-aircraft radar system in 1938-9.

Radio navigation ③

Radio navigation was developed during the **Second World War** as a means of directing bomber aircraft to their targets at night.

The German **Knickebein** ('bent leg') system, developed by Hans Plendl from 1934, was typical. It used a radio beam transmitted from Kleve, western Germany, to guide an aircraft to the target (Morse signals in the radio beam warned the pilot if he strayed off course). An interceptor beam from Bredstadt, northern Germany, signalled when the aircraft was over its target. Britain's Gee system and the US Loran were similar.

Modern radio navigation systems include the **VOR** system of radio beacons in commercial air corridors.

EYE ON THE SKY The VOR system, in combination with radar, enables air traffic to be controlled from a relatively small number of centres on the ground.

What is sonar? ❹

Sonar is a system for locating objects underwater.

◆ The **Asdic** system was developed by Allied scientists in 1917-18; it was renamed sonar by the US navy during the Second World War.

◆ **Active sonar** detects echoes from sound or ultrasound (extremely high frequency sound) generated by the system itself.

◆ **Passive sonar** detects sounds made by the target, for instance a ship or submarine.

◆ Modern systems are hull-mounted, or operate from an air-launched sonobuoy or unit suspended from a helicopter; passive sonar arrays are also installed on the seabed.

◆ The **echo sounder** found in many private boats uses sonar echoes from the seabed to determine water depth.

Acronyms ❺

Asdic Anti-Submarine Detection Investigation Committee (an apocryphal Allied scientific committee of the First World War)
Gee Grid navigation system
GPS Global Positioning System
Loran Long-range Radio Navigation System
Radar Radio Detection and Ranging
Sonar Sound Navigation and Ranging
VOR VHF (Very High Frequency) Omnidirectional Radio

TIMESCALE ❻

▶ **1904** German Christian Hülsmeyer builds a primitive radar system.

▶ **1908** Herman Anschütz-Kaempfe (Germany) invents the gyrocompass.

▶ **1911** Practical gyrocompasses built by Elmer Ambrose Sperry go on sale in the USA.

▶ **1917-18** Allied naval vessels are fitted with Asdic anti-submarine detection systems.

▶ **1922** Guglielmo Marconi proposes using radio wave echoes to detect ships.

▶ **1935** Tests in Britain prove radio wave echoes can detect distant aircraft.

▶ **1936** An early radar is fitted to the liner *Normandie* to detect icebergs.

▶ **1938** A chain of radar detectors to warn of aircraft attack is completed around Britain's south and east coasts.

▶ **1940-1** Germany's Knickebein, X-Gerät and Y-Gerät radio navigation systems guide bombers to their targets in Britain; all are soon jammed.

▶ **1964** The first satellite navigation system is created, using US navy satellites.

▶ **1978–95** The Navstar GPS system is set up, initially for the US military.

Global positioning ❼

The US Department of Defense's Navstar global positioning sytem (**GPS**) is the world's **most widely used navigational system**. Time signals regulated by atomic clock are transmitted by 24 satellites in precise orbits; at least four satellites are always simultaneously within range of any point on the Earth's surface. GPS receivers in boats, cars and even mobile phones compare signals from several satellites to plot a location accurate to a few metres.

PINPOINTED A GPS receiver calculates its distance from four or more satellites, as if it is on the surface of imaginary spheres centred on each of the satellites. The receiver's location is at the point where the spheres intercept.

⭐ 831

Spin control

The gyrocompass uses the principle that a freely spinning body tends to maintain a stable axis. If the axis of a free-mounted gyroscope is set to point north, it will continue to do so regardless of any other movement. Because they indicate true rather than magnetic north, gyrocompasses are widely used in marine navigation.

Telecommunications 1

Core facts ❶

◆ Telecommunication is any electric or electronic system for communicating at a distance. Both **cable** (telegraph or telephone) and **wireless** (radio telegraph or telephone) systems transmit almost instantaneously.
◆ The first telecommunication system, the **telegraph**, continued in use until 1999.

◆ The **telephone** was the first system to allow distance transmission of actual spoken words; telephones were rare until the 1920s, but then spread quickly around the world.
◆ The networks established for the telegraph and telephone systems form the basis of today's global telecommunications network.

Telegraph and telegrams ❷

Telegraphy is the sending of coded messages by electrical impulse over a network of wires. British scientists **Charles Wheatstone** and **William Cooke** invented the first electromagnetic telegraph system in 1837 (though the first telegraph message was sent in 1804 by **Francesco Salva** of Spain).

In 1838, **Samuel Morse's code** transformed the cumbersome early multi-wire telegraph system by allowing messages to be sent on a single line. Duplex and multiplex (simultaneous send and receive) systems were developed in 1872. From 1858, punched (ticker) tape allowed transmission at up to100 words per minute. **Teleprinters** (remote printers controlled by telegraph signal) were developed from 1855.

DELIVERED BY HAND Telegrams were written messages decoded from telegraph signals and delivered by special post.

The history of the telephone ❸

In 1876, the Scottish-American tutor of the deaf **Alexander Graham Bell** found a way of converting sound into an electrical impulse that could travel by wire. A thin steel plate, when vibrated by sound waves, caused a tiny, variable current in an electromagnet; when the current reached a second electromagnet at the end of a connecting wire, a similar metal plate vibrated in response, reproducing the original sounds. Within a few months, Bell had refined the system enough for words to be heard clearly.

Thomas Edison invented an improved mouthpiece and transmitter in 1877, and in 1878 the **first telephone exchange** opened with 21 subscribers. The **first long-distance calls**

ANYBODY THERE? Alexander Graham Bell inaugurates the New York-Chicago telephone line in 1892.

(from New York to Boston) were made in 1884. The **automatic exchange** (without operators to connect subscribers) was developed by Alvin Brown Strowger (USA) in 1889, and widely adopted from 1919. Fully **electronic exchanges** were not introduced until 1960.

BELL RINGER Bell's telephone was patented in 1876 and used by Queen Victoria.

What are radio waves? ❹

Radio waves are those **waves of electromagnetic energy** with the longest wavelengths and lowest frequencies (see also page 86).

◆ Other than **microwaves** (technically included among radio waves, and also used in telecommunication), radio waves include so-called ultra-high frequency (**UHF**) and very high frequency (**VHF**) waves, as well as the **short, medium and long waves** used in AM radio.

◆ The existence of radio waves was predicted by the Scottish physicist **James Clerk Maxwell** in the 1860s and proven experimentally by Germany's **Heinrich Hertz** in 1888.

◆ The Italian **Guglielmo Marconi** discovered in 1895 that pulses of radio waves could be detected at a distance by a suitable aerial or antenna, and the idea that this could be used for 'wireless' telecommunication was quickly developed.

◆ Marconi sent the first transatlantic wireless telegraph signal in 1901, from Cornwall in England to Newfoundland, off the coast of Canada.

★ 657

Citizen's band

Citizen's band (CB) radio is a **short-range** radio voice communication system using a combined transmitter and receiver (**transceiver**). The system of reserving a band of radio frequencies for CB use was developed in the USA in the 1940s. CB developed its own subculture and language in the late 1970s, but the growth of modern mobile communications has effectively killed it off.

Fax ❺

Facsimile transmission, or fax, was the first system for sending images by telegraph or telephone line.

◆ The British inventor **Frederick Blakewell** developed a primitive 'copying telegraph' in 1850, using messages written in non-conducting ink on tinfoil, a scanning electric stylus moved by pendulum and a clockwork cylinder transmitter.

◆ The first optical scanning system was developed by the German **Arthur Korn** in 1902, using photoelectric cells to distinguish light and dark in the image to be sent and a receiver that printed on photographic paper.

◆ Fax technology was gradually improved and standardised from the 1920s, leading to the development of **xerography** (forming an image by using static electricity to attract powdered pigment) in the 1950s. The modem in the early 1980s allowed the fast plain-paper faxing of documents of all types.

MAKING CONNECTIONS Telephone exchanges were once manned by large numbers of people whose job was to connect callers. Today's exchanges are fully automated.

Cables versus satellite ❻

Telecommunication by cable began with the telegraph system; the **first long-distance telegraph cable** was laid from Washington to Baltimore in 1844 and the first transatlantic cables in 1886. The first transatlantic telephone cable, TAT 1, was laid in 1956, and the first transatlantic fibre-optic cable in 1988.

Satellites provided a genuine alternative to cable from 1962, with the launch of **Telstar**. Signals from the ground were transmitted to the satellite, amplified and retransmitted to receivers on the other side of the Atlantic. Communications satellites are now used globally where cabling does not exist or is overloaded.

Where appropriate, cables are supplemented by microwave links between exchanges. With **fibre-optic cables**, these links also make possible the huge digital traffic of today's Internet communications.

WEIRD AND WONDERFUL ❼

Telegraphy a century ago was almost as efficient as modern email. Return telegrams and delivery-on-demand allowed several same-day responses. By 1900, the British were sending 90 million telegrams per year.

QUESTION NUMBER

The numbers or star
following the answers
refer to information
boxes on the right.

ANSWERS

281 **Finland** – founded 1865 as paper manufacturer

282 **7** – the letters are divided between keys 2 to 9

283 **Have a nice day**

284 **Short Message Service ❺**

287 **Will you go out with me?**

289 **Wireless Application Protocol ❺**

719 **Mobile phones ❶ ❺**

767 **Mobile phones ❶ ❺**

★ 805 **Pager** ★

899 **Television** – the others are telecom devices

936 **Cellular ❶ ❺**

938 **Third generation ❺**

939 **Integrated** – Integrated Services Digital Network

Telecommunications 2

❶

Telecommunications today

◆ Fast, accurate digital **encoding** of sound and other data underlies all modern telecommunications systems. The networks depend on a variety of technologies, including **fibre-optic** cable, satellites and **microwave** links. With the Internet, they play a key role in the world economy.

◆ The telephone network forms the basis of telecommunications systems. It was originally built to carry **analog** voice data (sound waves), but it now carries huge quantities of **digital** data from many different devices – telephones, fax machines, computers and a wide range of mobiles.

◆ **Mobile phone, or cellular, networks** play an increasingly important part in telecom systems. They divide the areas they cover into **'cells'**, each of which has its own transmitter-receiver base station or 'tower'. The base stations are all connected, by fibre-optic, microwave and other links, to the network's hub, the **network exchange**. When mobile phones are switched on, they transmit **continuous signals** which allow computers at the base stations to keep track of them.

In the illustration below, copper wire and fibre-optic cable links are shown in green, microwave links in pink.

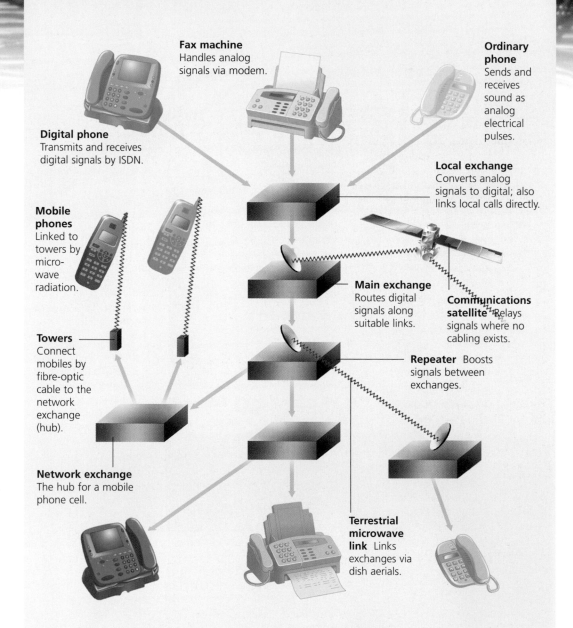

Fax machine Handles analog signals via modem.

Ordinary phone Sends and receives sound as analog electrical pulses.

Digital phone Transmits and receives digital signals by ISDN.

Local exchange Converts analog signals to digital; also links local calls directly.

Mobile phones Linked to towers by microwave radiation.

Main exchange Routes digital signals along suitable links.

Communications satellite Relays signals where no cabling exists.

Towers Connect mobiles by fibre-optic cable to the network exchange (hub).

Repeater Boosts signals between exchanges.

Network exchange The hub for a mobile phone cell.

Terrestrial microwave link Links exchanges via dish aerials.

Microwave link

Microwaves are a form of **electromagnetic radiation** (see page 86). They are good for transmitting information because microwave energy penetrates most weather conditions, including cloud, snow and haze. Microwaves play a key part at two points in the modern telecommunications system:

◆ Cellular phones communicate with the local cell's **hub** transmitter-receiver via microwave at a particular frequency. They switch automatically to a different frequency if the user moves from one cell to another.

◆ Microwave links also often connect local and main exchanges in the network itself, via **dish aerials** arranged along a line of sight.

WEIRD AND WONDERFUL

Medical evidence suggests that teletexting on mobiles may actually be changing the **physiology** of the **human hand**. It is helping to make the thumb the most versatile digit, in place of the forefinger.

Digital conversation

Data of any sort can be handled by today's telecommunications network, provided that it can be encoded in digital form – that is, as **binary digits**.

Telephone sound is encoded by an **analogue-digital converter (ADC)** which samples (or measures) the amplitude (height) of a sound wave 8000 times per second, on a scale of 0 to 255. Each measurement is transmitted as a series of pulses representing binary digits.

On receipt, the original sound is regenerated by a **digital-audio converter (DAC)**.

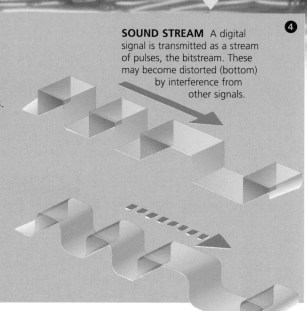

SOUND STREAM A digital signal is transmitted as a stream of pulses, the bitstream. These may become distorted (bottom) by interference from other signals.

TIMESCALE

▶ **1837** In Britain, Charles Wheatstone and William Cooke pioneer the electric telegraph.
▶ **1858** The first transatlantic telegraph cable is laid (Ireland-Newfoundland)
▶ **1872** Multiplex telegraph transmission is developed by US inventor Thomas Edison.
▶ **1876** Alexander Graham Bell invents the telephone.
▶ **1878** The first telephone exchange is opened, at New Haven, Connecticut, USA.
▶ **1901** Guglielmo Marconi transmits the first transatlantic radio signal.
▶ **1921** Police in Detroit, USA, start using mobile pagers.
▶ **1947** Bell Labs in the USA invent cellular mobile phones.
▶ **1958** The modem is invented.
▶ **1961** The first telecoms satellites are launched.
▶ **1970** Fibre-optic cable is first demonstrated.
▶ **1978** The first cellular phone systems are set up in Chicago and Tokyo.
▶ **1988** The Integrated Services Digital Network (ISDN) system is launched in Japan.
▶ **1993** The first digital mobile network is set up, in Los Angeles.
▶ **2000** 552 million mobile phones are in use worldwide.

The mobile explosion

Telecommunications have been dominated in recent years by the explosive rise of mobile phone usage.

◆ The first generation (1G) systems in the 1980s used 'cellular mobile radio telephones' – large handsets using **analog** signals.

◆ In the 1990s, second generation (2G) systems featured increasingly miniaturised **digital** handsets. 2G technology continued to improve, notably in the use of **WAP (Wireless Application Protocol)** to deliver text and multimedia, leading to the so-called 2.5G standard.

◆ Third generation (3G) systems, now appearing, feature **high bandwidth**, allowing Web surfing, videoconferencing and so on, and **roaming** capability. They can be used on networks throughout Europe, Japan and North America.

◆ The spread of mobile usage is revealed by the fact that UK users sent over 1 billion text messages in June 2002 alone.

SEE AND BE SEEN Third generation videophones not only show the person you are talking to, but also show the picture the other person is getting of you. They can also send photos and sound in 'text' messages.

★ 805

Page me

A pager is a dedicated device for communicating a simple message using a **specific radio frequency**.

The first pagers simply alerted the user that a message was waiting. Second-generation pagers feature an LCD screen for messages.

The system of alerting a pager by telephone was patented in 1949 by American **Al Gross**.

Radio and TV

Core facts ❶

◆ Radio and television are both inventions of the 20th century; from the 1920s onwards, their social impact has been enormous.
◆ Broadcast radio and television are both transmitted using **radio waves**.
◆ In both media, carrier signals of particular **frequencies** are assigned to individual stations by national and international agreement, to prevent **interference**.
◆ **Cable** and **satellite** increasingly provide alternative ways of transmitting television, and **digital** technology is revolutionising both sound and image quality; radio content is also now widely available via the **Internet**.

TIMESCALE ❷

▶ **1888** Radio waves are demonstrated by German physicist Heinrich Hertz.
▶ **1895** Guglielmo Marconi (Italy) and Alexander Popov (Russia) independently transmit radio signals.
▶ **1906** Reginald Fessenden makes the first radio transmission of voice and music (Massachusetts, USA).
▶ **c.1909** San Francisco station KCBS makes the first scheduled radio broadcasts.
▶ **1922** The BBC is founded.
▶ **1923** Russian-born Vladimir Zworykin invents the iconoscope, the first electronic television camera.
▶ **1926** British inventor John Logie Baird demonstrates his mechanical television.
▶ **1927** NBC and CBS begin radio broadcasts in the USA.
▶ **1928** General Electric (USA) and the BBC (Britain) begin television transmission. Regular BBC service begins in 1936, from a transmitter at Alexandra Palace, London.
▶ **1940** FM (frequency modulated) radio broadcasting begins in USA.
▶ **1951** Regular colour television transmission begins in the USA.
▶ **1955** Pocket transistor radios go on sale in the USA.
▶ **1960** Stereo radio broadcasting in FM begins in the USA.
▶ **1962** Telstar, the first television satellite, is launched.
▶ **1975** The first satellite-cable television networks are established in the USA.
▶ **1991** British inventor Trevor Baylis unveils his wind-up clockwork radio.
▶ **1994** Digital television is first broadcast by satellite.
▶ **1995** Digital radio is introduced in the UK and Canada.
▶ **1996** The first Web radio broadcasts are made, using the Internet.
▶ **1998** Digital television broadcasts begin from terrestrial transmitters.

⭐ 660 Web radio

Web radio programmes come to your computer via the Internet, rather than by radio transmission. **Streaming audio formats** allow sound files to start playing even as the file is being downloaded. Many radio stations now offer Web radio alongside broadcast services.

Television's first face ❸

In 1926, in an attic above a restaurant in Soho, London, Scottish inventor John Logie Baird gave the world's first successful demonstration of television. He scanned and re-created the face of office boy William Taynton.

How radio works ❹

Sound waves are detected by a **microphone** and converted to electrical impulses. These are superimposed on a radio wave (**carrier wave**) generated by an **oscillator**. The combined (**modulated**) radio wave is **broadcast** by the large aerial on a transmitter mast.

The broadcast wave is detected by a radio **receiver** tuned to the carrier's frequency. In the receiver, the electrical impulses of the sound signal are separated from the carrier by a **demodulator**, amplified and converted back to audible sound waves by a **loudspeaker**.

In amplitude modulation (**AM**), the amplitude (power) of the carrier wave is varied to match the voltage of the sound signal. In frequency modulation (**FM**), the frequency of the carrier wave is varied to match the sound signal; this reduces interference. The clearest signal of all is produced by **digital audio broadcasting** (**DAB**), in which the sound wave is converted into a sequence of digital bits; these are decoded by a microchip in the receiver.

1930s TV A vision-only 5-inch Pye television receiver, dating from 1938.

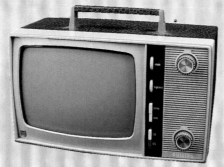

1960s TV A portable Philips television receiver, dating from the 1960s.

The box has changed ❺

◆ The earliest generation (1930s and 1940s) of television sets, like radio receivers of the time, were set in **wooden cases** that matched contemporary cabinet furniture. Screens were circular and small (typically 5, 9 or 12 inches in diameter).

◆ In the 1950s and 1960s, **plastics** transformed the exterior of the set while technological advances reduced its size, and improvements in **cathode-ray tube** (CRT) manufacture brought larger, rectangular screens. The vast majority of sets remained black-and-white.

◆ Further technological advances in the 1970s led to miniature, **portable** sets, improved sound and still larger, flatter screens, and to the widespread adoption of **colour**.

◆ From the 1980s onwards **video recording**, **stereo sound**, high-definition and wide-format screens, and full '**home cinema**' systems have been accompanied by a high-tech look that proclaims the television's technical sophistication.

◆ 'Flat-screen' **plasma display panels** (PDPs) for TV sets are increasingly replacing the bulky cathode-ray tubes needed for conventional electron guns. In PDPs, a grid of electrodes carries an electrical current directly to individual cells called pixels. Red, blue and green subpixels within each light up selectively, creating the image. PDPs offer much higher resolution than CRTs – about 1100 lines at present.

1980s TV A Sony Watchman 'Voyager' pocket television receiver, dating from 1982.

2000 TV A Sony plasma TV screen, dating from 2002.

FLAT-SCREEN TECHNOLOGY In the 1980s manufacturers began producing television sets with screens only a few centimetres deep.

Vertical electrode

Pixel

Horizontal electrode

How TV works ❻

Images from a TV camera are split into red, blue and green, and scanned electronically into 525 or 625 horizontal lines. The colour signals are combined with an **audio** signal (for sound), a **synchronisation** signal and a radio wave for transmission. The transmitted signal is received by a TV set, which re-creates the image using **electron guns** that scan the screen in lines, causing red, blue and green **phosphor stripes** to light up.

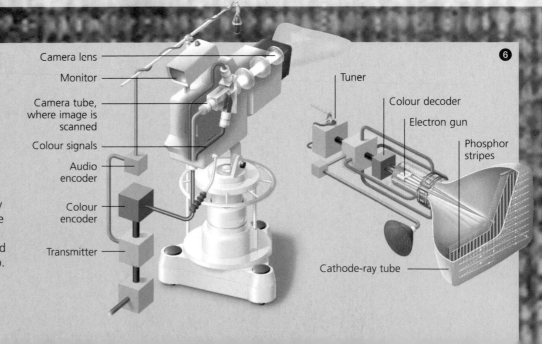

Camera lens

Monitor

Camera tube, where image is scanned

Colour signals

Audio encoder

Colour encoder

Transmitter

Tuner

Colour decoder

Electron gun

Phosphor stripes

Cathode-ray tube

QUESTION NUMBER

The numbers or star following the answers refer to information boxes on the right.

Photography and film

Core facts ❶

◆ British astronomer Sir John Herschel coined the term photography in the 1840s from the Greek *photos* (light) and *graphein* (to write).
◆ A camera records a **pattern of light** either on light-sensitive film or as digital information via a charge-coupled device (CCD).

◆ A **motion picture** ('movie') camera records a rapid succession of still images, which appear to show movement when played back.
◆ How a particular image is recorded depends on accurate control of the **amount of light** (exposure) reaching the film or CCD.

The camera

A camera typically includes a **viewfinder**, a diaphragm for **aperture** control and a **lens** system for focusing the image. A **telephoto** lens brings distant objects closer, but compresses perspective; a **wide-angle** lens gives a wide view but exaggerates perspective. A **zoom** lens is variable between these two.

The first truly portable cameras were made from 1888 by George Eastman's Kodak company. The '**Brownie**' box camera was introduced in 1900. Oskar Barnack in Germany invented the first miniature camera in 1912. The 35mm **single-lens reflex** (SLR) camera was first produced in 1935. Mass market **digital cameras** first went on sale in 1996.

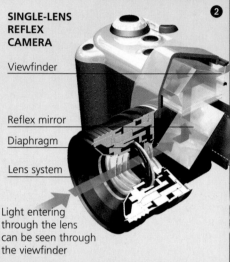

❷

SINGLE-LENS REFLEX CAMERA

Viewfinder

Reflex mirror

Diaphragm

Lens system

Light entering through the lens can be seen through the viewfinder

SPLIT-SECOND TIMING A high-speed photograph captures the moment a .22 calibre bullet smashes through a raw egg.

Types of camera ❸

Plate camera	Single-lens reflex (SLR) camera	Compact camera	Polaroid camera	Advanced photo system (APS) camera	Digital camera
Large-format camera, similar to 19th-century examples. Exposes single-sheet film 13 x 10cm (5 x 4in) in size. Used by professionals for high-quality, generally static, images.	Versatile camera with a prism system that allows the user to compose the picture directly through the lens. Used with 35mm or 5.5 x 5.5cm (2¼ x 2¼in) film.	Miniaturised camera with a separate viewfinder and a lens that cannot be changed. Generally used for informal photography, with 35mm or smaller film.	Compact camera using single-sheet film that includes processing chemicals. The camera develops the exposed film automatically, producing a print within a few minutes.	Compact camera with sophisticated electronics, using film in a cassette. Built-in zoom lenses and electronic systems allow wide-screen or panoramic images.	Images are recorded by light-sensitive electronic cells (in the CCD), rather than film. These digital images are viewed on a screen, not a viewfinder, and are uploaded to a computer for printing.

Technical terms

◆ **Aperture** The size of the opening through which light reaches the film, measured in f-numbers or f-stops.
◆ **Depth of field** The distance over which objects remain in sharp focus. A wider aperture (f1.4, say) gives a shorter depth – only objects near the point of focus of the lens are sharp. A smaller aperture (f22, say) gives a greater depth.

◆ **Film speed** Measured in ISO numbers, it shows how fast a film reacts to light. Fast films (ISO 400 and upwards) require less light, but give a more 'grainy' image.
◆ **Shutter speed** The length of time that the shutter remains open. Faster speeds (1/100th of a second or less) freeze moving subjects. Slower speeds allow longer exposure (in poor light, say).

★ 244

Cartoons

The first cartoons, or **animated films**, appeared around 1905. Walt Disney's creation Mickey Mouse appeared in *Steamboat Willie* in 1928. *Snow White and the Seven Dwarves* was his first feature-length animation in 1937. Nowadays, many film-makers use sophisticated computer animation.

TIMESCALE

▶ *c.1725* Johann Schultze in Germany discovers that silver salts become darkened by exposure to light.
▶ **1839** Frenchman Louis Daguerre invents the Daguerrotype, a system for reproducing single, high-quality photographic images. Englishman W.H. Fox Talbot invents the calotype, allowing several 'positives' to be developed from one 'negative'.
▶ **1872** English photographer Eadweard Muybridge uses 24 cameras to record the gait of a galloping horse.
▶ **1873** Silver bromide paper is introduced for camera film.
▶ **1890-4** In the USA, Thomas Edison invents the Kinetograph camera, the earliest successful movie camera.
▶ **1895** The Lumière brothers open the first public cinema, in Paris.
▶ **1927** *The Jazz Singer*, the first major film including dialogue in synchronised sound, is released.

FIRST PHOTO In 1825, French chemist Nicéphore Niepce used chemicals that react with light to make this image of a 17th-century print.

▶ **1932** The Technicolor colour film process is first used, in an animation.
▶ **1969** The charge coupled device (CCD) is invented, beginning the development of digital photography.
▶ **1982** The film *Tron* makes the first effective use of computer-generated animation.
▶ **1990s onwards** Digital graphics play a major part in film-making: for example, in *Toy Story* (1995, animation) and *Gladiator* (2000, cloning for crowd scenes).

Moving pictures

The motion picture camera has a motor-driven **film transport system** that enables it to expose successive images on a reel of 8, 16, 35 or 70mm film. The film is spooled from the camera's forward magazine into an **exposure chamber**, where it is gripped by a mechanical claw and locked in place for exposure. The claw then advances the film by one step (frame) for the next exposure, at a rate of 24 or 30 **frames per second** (fps). The exposed film then spools into the rear magazine.

A **projector** uses a similar mechanism to project each frame of the processed film in succession onto a screen, using a powerful lamp.

PIONEER PROJECTOR Brothers Auguste and Louis Lumière's *cinematographe*. The projectionist turned the handle to pull the film down from the top spool, through the projector and into a box beneath.

The stages of film production

Film production has three phases:
◆ **Preproduction** The studio approves a film idea and appoints a producer to manage the project. The producer and studio arrange finance. The producer hires a director, actors and technical crews. The writers revise the filmscript.
◆ **Production** The set designers create the sets. The director directs the actors and camera operators as scenes are shot. Dialogue, sounds and special effects are added.
◆ **Postproduction** The film editor cuts and assembles the raw film footage, giving the film its final shape. Music is added. Test screenings are arranged for preview audiences. The film is released to cinemas.

Recorded sound and pictures

Core facts ❶

◆ Recording sound became possible in the late 19th century, and was a **major entertainment industry** by the first decade of the 20th.
◆ Sound-recording for domestic playback uses an electro-mechanical system (**records**), an electro-magnetic system (**tape** or **cassette tape**) or a digital system (**CDs**).

◆ Before modern digital systems, **film sound** used an optical system with the soundtrack recorded on the film itself.
◆ **Video recording** uses electro-magnetic technology to store both sound and television images on large-format cassettes; it is steadily being replaced by **digital recording** on DVDs.

First sound recording ❷

In 1877, US inventor Thomas Edison was working on a system for converting Morse code signals to marks on paper. Out of curiosity, he tested the system for the **human voice**, allegedly by reciting 'Mary Had a Little Lamb' into a horn leading to the diaphragm from a telephone mouthpiece. A **stylus** attached to the diaphragm recorded its vibrations as impressions on a **revolving cylinder** of dampened paper (later he used aluminium foil). When he wound the cylinder back again, the impressions made the stylus and diaphragm vibrate – and, faintly, he heard **his own voice** repeating the nursery rhyme from the horn.

CAPTURING THE VOICE
A replica of Edison's first phonograph.

Silent movies to talkies ❸

The first movies were accompanied by live music, but from 1907 the development of amplification made **film sound** feasible. Warner Brothers used the Vitaphone phonograph disc system for *The Jazz Singer* (1927), the first film with **synchronised speech**. By the 1950s, the RKO/RCA Photophone sound-on-film system had become standard. It used a photocell to read an **optical soundtrack** – with varying width – on the film itself.

FIRST TALKIE   Al Jolson in *The Jazz Singer*. The soundtrack had more songs than dialogue.

Technology of record, tape and CD ❹

The three recording media that have successively dominated popular sound recording and playback are the record (disc), the cassette tape and the CD. Of these, the record allowed **playback only**, whereas tape and CD have both also been marketed for **domestic recording** as well as playback.

RECORD

TAPE OR CASSETTE

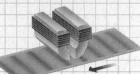

AUDIO CD

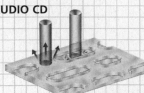

Sound information is carried by oscillations in a **spiral groove**. As the stylus runs along this, electromagnets produce electric currents, which the loudspeakers convert to sound waves.

Sound information is carried by **varying magnetisation** of the tape's metallic coating. As the tape runs past the playback head, these variations produce electric signals from electromagnets.

Digital data (see page 95) is stored as **pits** and **lands** (unpitted areas) in a spiral pattern. During playback, a laser follows the spiral, and a digital-analogue converter decodes the data as sounds.

Video and DVD ❺

◆ Magnetic tape was first used to record **video** (television signals) in 1951. Video data is recorded magnetically in the same way as audio information, but multiple heads are used to record (and play back) multiple streams of data. Sony introduced the high-quality **Betamax** video format in 1975, but JVC's lower-quality **VHS** format became the standard.

◆ The **Digital Versatile Disc (DVD)** was introduced in 1996 in Japan. A double-layer recording surface on the disc means that up to 25 times as much data can be stored on a DVD than on the earlier video or laser CD disc – enough for an entire feature film at higher quality than video.

★ **221**

The jukebox

In 1889 American Louis Glass installed a coin-operated Edison phonograph in a San Francisco saloon. In 1905 a jukebox with 24 cylinder recordings was invented. But the first true, **all-electric jukebox** came in 1927. The machines became popular with the manufacture of the **Wurlitzer** multiselection example in 1934.

Stands for ... ❻

CBS Columbia Broadcasting System (USA)
EMI Electric & Musical Industries Ltd (UK)
JVC Japanese Victor Company (part of the Matsushita group)
RCA Radio Corporation of America
RKO Radio-Keith-Orpheum Corporation (USA)

TIMESCALE ❼

▶ **1878** Thomas Edison forms the Edison Speaking Phonograph Company.
▶ **1886** Americans Charles Turner and Chichester Bell introduce wax cylinders as a recording medium.
▶ **1888** In the USA, Emile Berliner invents the record disc and builds the first gramophone.
▶ **1898** Danish inventor Valdemar Poulson pioneers the use of magnetism to record sound.
▶ **1925** The microphone is developed for making sound recordings.
▶ **1931** Alan Blumlein of EMI invents binaural (stereo) recording.
▶ **1935** The first tape recorder (the Magnetophon) is developed in Germany by Fritz Pfleumer and the BASF and AEG companies.

▶ **1941** Stereo sound is first used in a cinema.
▶ **1948** CBS introduces plastic (vinyl) for discs, making possible the long-playing (LP) record.
▶ **1963** The Dutch company Philips introduces the tape cassette.
▶ **1965** Sony of Japan produces the first consumer videotape recorder.
▶ **1969** In the USA, Ray Dolby invents the Dolby noise-reduction system.
▶ **1979** Akio Morita of Sony invents the Walkman portable personal cassette player.
▶ **1982** Sony and Philips develop the compact disc (CD).
▶ **1992** Sony introduces the digital re-recordable Minidisc.
▶ **2001** Domestic DVD recorders and recordable (DVD-R) discs go on sale.

Make your own movies ❽

Simple **cine cameras** and projectors, using 8 mm film, were made from 1932 onwards for home filmmakers. The development of video cassette recorders in the 1970s and, in 1980, the first domestic video camera (**camcorder**) made the creation of home videotapes possible.

Advances in computer technology in the mid 1980s meant that data could be recorded onto compact discs. The commercial breakthrough for home 'filmmaking' came in the late 1990s, with the release of a wide range of **digital camcorders** and digital editing software for both images and sound.

TV PLAYBACK Images shot on a digital video camera can be played back in the normal way, using a video recorder.

COMPUTER PLAYBACK Digital video images can also be fed into a computer, edited and stored on a compact disc.

Printing

Core facts ❶

◆ Printing is the creation of multiple copies of text and images. It depends on several processes.
◆ Uniform lettershapes (**typefaces**) are arranged into pages of text for the printing plate, a process known as **typesetting**.
◆ **Illustrations** are converted into lines using woodcuts or engraved metal plates (mezzotints),

or into dots using a half-tone screen.
◆ Ink is applied to the printing plate in a **press**, and the plate is brought into precise and even contact with the paper.
◆ Gutenberg's press could print about 16 copies an hour. Modern newspaper presses can print up to 90000 an hour.

CMYK ❷

Because printing inks cannot mix on the page and only a manageable number of inks can be used at once, **colour printing** is mostly done using the four-colour process method. Four inks in different combinations – **cyan** (blue), **magenta** (red), **yellow** and '**key**' (black) – provide the widest possible range of final colours. The colour image is photographically broken down (screened) into patterns of fine, different-sized dots reflecting the varying intensity of the CMYK inks in the colours required. Each ink has its own pattern (**separation**), which is printed separately in turn; adjacent dots of colour mix visually in the reader's eye, creating a full range of colours and intensities.

FOUR-COLOUR PROCESS In colour printing, four coloured inks – yellow, magenta, cyan and key (black) – are applied, one after the other, to the sheet of paper. Together, they give the full range of coloured tones.

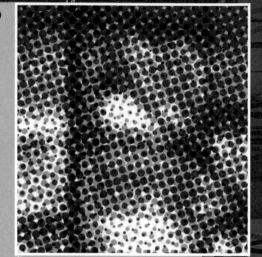

PRINTED PICTURE This small section of the background photograph of the *Financial Times* press in London has been blown up to show the thousands of colour dots that make up the image.

Y
MY
CMY
CMYK

Typefaces ❸

Variety in **typefaces** (fonts) played a part in printing from the beginning. **Gutenberg** used a heavy Gothic face for his 42-line *Bible* of 1455. Many thousands of fonts have been designed since, including fonts with **serifs**, the fine lines finishing off letters (e.g. Baskerville and Bodoni) and fonts (such as this one) without serifs or **sans serif**. Most fonts include a variety of **weights** (bold, semi-bold, medium, light) as well as **styles** (italic, small capitals).

HUMANIST
From the 1470s
Wide, Roman-style letters with little weight difference between horizontals and verticals

OLD STYLE
Caslon (1734)
Narrower letters, medium weight contrast between horizontals and verticals.

623

Diamond Sutra

The earliest surviving **printed work** (a scroll rather than a book) is a version of the Buddhist scripture known as the *Diamond Sutra*, printed in China in 868. It was printed using carved 75 cm (2¹/₂ ft) long **woodblocks** and includes both text and pictures. A 4.5 m (16 ft) section survives in the British Museum, London.

TIMESCALE

▶ **9th century AD** Woodblock printing begins in China.
▶ **1450** Gutenberg begins commercial printing in Europe.
▶ **1461** Albrecht Pfister prints the first illustrated book.
▶ **1476** William Caxton sets up the first press in England.
▶ **1477** Gravure printing is invented in Europe.
▶ **1719** Jakob Le Blon invents full-colour printing and the process for engraving metal.
▶ **1798** Alois Senefelder invents lithography.
▶ **1852** William Fox Talbot invents the half-tone process for printing images.
▶ **1865** Richard Hoe patents a high-speed rotary press.
▶ **1880s** The introduction of the Linotype and Monotype machines greatly increases typesetting efficiency.
▶ **1939** William Huebner invents a photocomposition machine, which transfers type photographically to the plate.
▶ **1984** Desktop publishing (DTP) software is introduced.
▶ **1990s** Digital presses and print-production software make plateless, completely electronic printing feasible.

Gutenberg ❹

Johann Gutenberg, a silversmith from Mainz, Germany, is regarded as the inventor of modern printing, around 1450. Gutenberg and others had experimented with **cast metal letters** since the 1430s, but he and his associates Johann Fust and Peter Schöffer made two crucial advances: they used steel punches to create uniform letter moulds, and they adapted a screw-worked textile press to force the sheets of paper evenly against the inked letters. Gutenberg's *Donatus Latin Grammar* (1451) is the first known book printed in this way.

PRINTED BIBLE The illuminated first page of Gutenberg's 42-line *Bible*, printed in 1455.

Printing methods

Historically, there are three main types of printing:

◆ **Letterpress** The earliest form, in which the type or images are raised above the **printing plate** surface. The ink roller touches only these areas, so only they print when paper is pressed onto the plate. Letterpress survives only as an artform, for very specialist printing.

◆ **Photogravure** The opposite of the letterpress process. The type or image is **etched** into the plate using a photographic process. Ink is applied to the whole surface, but it is then wiped away from the raised areas; it is retained by the etched areas, and absorbed from these by the paper. Gravure is mostly used for high-quality art printing and for printing on non-paper surfaces. ❻

◆ **Lithography** The image (type and illustrations) is made into film and transferred photographically to the plate, which is chemically treated so that ink sticks only to the image areas. In **offset lithography**, the inked image is transferred (offset) to a rubber roller, the **blanket**, which then rolls across the paper. Offset lithography is now the most versatile and widely used type of printing.

Typesetting transformed ❼

Until the late 19th century, printworkers painstakingly arranged individual metal letters in a frame to make up pages for printing. The **Linotype** and **Monotype** machines revolutionised setting by casting whole lines of text in hot metal as the setter typed at a keyboard. With **DTP**, whole pages laid out on computer are converted photographically to printing film in a phototypesetter, or even input directly to a digital press.

Cc Dd Ee Ff Gg Hh

TRANSITIONAL Baskerville (1757) Crisp horizontals and verticals showing medium-to-high weight contrast.

MODERN Bodoni (1787) Inspired by Classical style, with strong weight contrasts between verticals and horizontals.

EGYPTIAN OR SLAB SERIF Rockwell (1934) Machine-crafted, with little weight contrast and heavy, prominent serifs.

SANS SERIF Gill Sans (1929) Generally wide letters without serifs, sometimes very heavy in weight.

SCRIPT Kuenstler Script Intended to reflect handwriting; some mimic comic-style hand lettering.

OPTICAL CHARACTER RECOGNITION (OCR) Geneva Designed so that it is easy to scan and record digitally.

The laser

Core facts ❶

◆ A laser (short for 'light amplification by stimulated emission of radiation') is a device for producing a very intense **beam of light**.
◆ A laser beam can travel over great distances and be focused to produce **high energy**. A beam can be produced from many different substances, including solids, liquids and gases – artificial rubies are often used – and from some chemical reactions.
◆ **'Natural' lasers** have been detected in space – for example, part of the Orion nebula produces laser light at near-infrared frequency.

From maser to laser ❷

Working from a theory originally put forward by Einstein, US physicist **Charles Townes** built a forerunner of the laser in 1953. Known as the **maser**, it produced a beam of microwaves, rather than light. Townes and his colleague **Arthur Schawlow** then suggested that the same principle could be applied to visible light. Another US physicist, **Theodore Maiman**, built the first laser in 1960. But technical difficulties and the decline of military funding for research as the Cold War waned meant that at first few practical applications were found. Most current applications were developed from 1990 onwards.

How a laser works ❸

An atom excited by heat, light or collision with other atoms emits 'packets' (or photons) of **light energy**. If it is irradiated with light of a particular wavelength, it will emit the light energy in phase with that wavelength, thus **amplifying** the light. The light is then 'bounced' between two mirrors – one normal, the other semi-transparent – at opposite ends of a tube, intensifying with each pass, until it emerges through the semi-transparent mirror as a **laser beam**.

MAKING THE BEAM The material in the laser tube is energised – here with an electric charge – and the 'packets' of light energy are bounced back and forth inside the tube.

★ 761

Sci-fi

Laser light, with its ability to 'burn' through hard materials, has often appeared in science fiction, most famously in the **light sabres** wielded by the Jedi knights in the *Star Wars* films. Curiously, one of the best descriptions of a laser-like weapon comes from H.G. Wells's *The War of the Worlds*, published in 1898 well before the laser had even been conceived.

Holograms ❹

A hologram is an image of an object, made using a laser beam split into two. One beam is directed at a photographic plate; the other is reflected off the object onto the plate. When light of the same frequency as the laser later illuminates the hologram, a **three-dimensional image** of the object is created.

HOLOGRAM'S INVENTOR A holographic image of the process's inventor, British physicist Dennis Gabor.

LASER SHOW At night, lasers are often used to create light displays, as here at the Glastonbury pop festival.

WEIRD AND WONDERFUL 6 The precision of a laser beam is such that it can be used to detect **precise vibrations** in window glass caused by sound waves. Laser surveillance equipment could thus record a conversation inside a room.

PRECISION BEAM Eye surgeons use lasers to re-attach detached retinas, mend broken blood vessels and correct eyesight defects.

Lasers in action 5

Bar codes Lasers are used to scan the bar codes printed on most packaged products. Their accuracy means that the code is usually captured by even the most cursory pass of the product through the scanner.

CDs Lasers are used both to burn (record) and to play back all forms of digital storage, including CDs and DVDs. High volumes of digital data can be encoded in a small area.

Industry Lasers provide precise alignment for construction (for example, in drilling and tunnelling); to drill accurate holes through even the toughest substances (such as diamonds); for small-scale cutting and welding (as in microcircuits); and to inspect optical and other equipment.

Laser printing Data from a computer turns a laser on and off as it scans a charged drum in the printer. Toner ('ink') is attracted only to those areas of the drum not illuminated by the laser. So only those areas are transferred to the paper.

Light shows Low-power lasers create spectacular light shows, particularly when combined with mirrors and atmospheric effects like dry-ice clouds.

Measurement The precision of laser light makes it suitable for extremely accurate measurements. For example, a laser reflected from a mirror placed on the Moon has measured the distance from the Earth to the Moon to within a few centimetres.

Optical fibres Because of its high frequency, light from lasers can be modulated to carry a huge amount of information. Laser light shielded within optical-fibre cables is key to modern high-bandwidth communications.

Security cards The use of holograms on security cards makes them more difficult to counterfeit.

Surgery A laser scalpel is sharp enough to vaporise parts of an individual cell in microsurgery, and its heat seals small blood vessels, reducing bleeding. Lasers are used in eye surgery, both on the retina and to alter the shape of the lens to correct short-sightedness.

Weapons Lasers are used to target smart bombs and other weapons. Their most likely use as destructive weapons in their own right is probably in satellites, where atmospheric interference is not a factor.

CUTTING EDGE Lasers are widely used to cut and weld metal piping.

Mass production

Core facts ❶

◆ Mass production is the production of long runs of **standardised** goods for a mass market.
◆ It has its origins in the Industrial Revolution of the 18th century, but only became truly influential in the 20th, with the development of the **production line**.

◆ The basic principle is that workers or machines each performing just one task are **more productive** than the same number of workers each performing all the tasks in a process.
◆ **Computers** and **robots** have further transformed production since the 1970s.

The production line ❷

The growth of industry brought increasing **specialisation** and the **division of labour**, whereby particular employees specialised in particular tasks. The production line extended these principles by moving the **work to the worker**.
  Henry Ford got the idea from **conveyor belts** he had seen in Chicago's meat-packing plants. From 1913, he put this into effect in the motor industry to move car chassis automatically between work stations. This reduced the time workers spent moving

from car to car, so they completed their specialised tasks more quickly as they had fewer movements to make.
  In modern production lines, human workers are increasingly replaced by **robots**, and the progress of each workpiece along the line is controlled by **computer**.

STACKED HIGH Furniture-makers, such as the British company Ercol, apply mass production techniques to the manufacture of wooden chairs and tables.

CAR MAKERS Robots weld car bodies on an assembly line in a Mazda factory in Hiroshima, Japan.

Robotics ❸

The word **robot** was coined by the Czech playwright Karel Capek in his play *RUR (Rossum's Universal Robots)*, first performed in London in 1920. It is derived from the Czech *robota*, 'slave labourer'.
  In 1942, science fiction writer Isaac Asimov published a short story, 'Runaround', in which he proposed three 'laws of robotics' to be programmed into a robot's 'brain'. The first law stated that a robot may not injure a human being, or through inaction allow any human being to come to harm.
  Besides manufacturing, modern robots have a role in **space exploration** (for example, NASA's 1997 Mars *Sojourner*), and in surgery, entertainment and warfare. In 1999, the Sony Corporation introduced Aibo, a **toy robotic dog**. Updated several times since then, it incorporates a highly developed Artificial Intelligence system and recognises 75 spoken commands.

Make-believe mechanoids ❹

From their origins in Karel Capek's play and Fritz Lang's classic futuristic film *Metropolis* (1927), robots that mimic human appearance ('androids') or display human or animal behaviour have been popular. They include: **R2D2** and **C3PO** (right) in George Lucas's *Star Wars* films; **Marvin the Paranoid Android**, a depressed robot in Douglas Adams' *The Hitchhiker's Guide to the Galaxy* (1979); and **Kryten**, a rebellious butler in the BBC comedy sci-fi series, *Red Dwarf*. Kryten's name refers to J.M. Barrie's castaway butler in his 1902 play *The Admirable Crichton*.

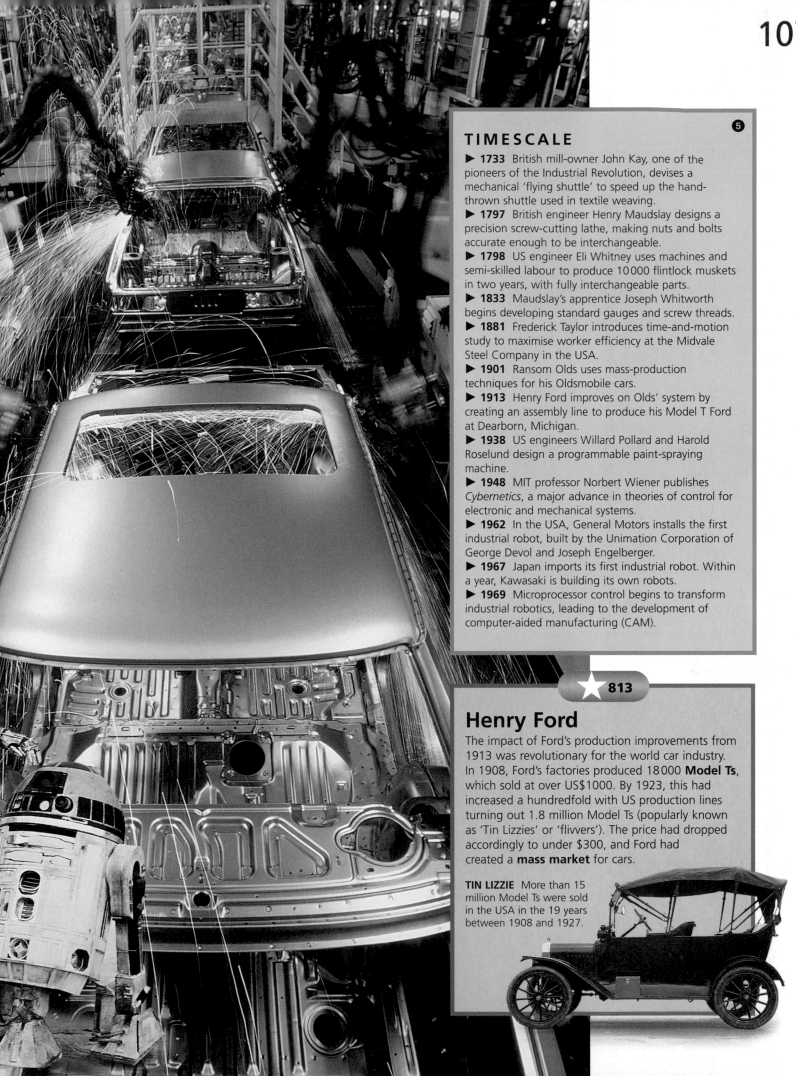

TIMESCALE

❺

▶ **1733** British mill-owner John Kay, one of the pioneers of the Industrial Revolution, devises a mechanical 'flying shuttle' to speed up the hand-thrown shuttle used in textile weaving.

▶ **1797** British engineer Henry Maudslay designs a precision screw-cutting lathe, making nuts and bolts accurate enough to be interchangeable.

▶ **1798** US engineer Eli Whitney uses machines and semi-skilled labour to produce 10000 flintlock muskets in two years, with fully interchangeable parts.

▶ **1833** Maudslay's apprentice Joseph Whitworth begins developing standard gauges and screw threads.

▶ **1881** Frederick Taylor introduces time-and-motion study to maximise worker efficiency at the Midvale Steel Company in the USA.

▶ **1901** Ransom Olds uses mass-production techniques for his Oldsmobile cars.

▶ **1913** Henry Ford improves on Olds' system by creating an assembly line to produce his Model T Ford at Dearborn, Michigan.

▶ **1938** US engineers Willard Pollard and Harold Roselund design a programmable paint-spraying machine.

▶ **1948** MIT professor Norbert Wiener publishes *Cybernetics*, a major advance in theories of control for electronic and mechanical systems.

▶ **1962** In the USA, General Motors installs the first industrial robot, built by the Unimation Corporation of George Devol and Joseph Engelberger.

▶ **1967** Japan imports its first industrial robot. Within a year, Kawasaki is building its own robots.

▶ **1969** Microprocessor control begins to transform industrial robotics, leading to the development of computer-aided manufacturing (CAM).

★ 813

Henry Ford

The impact of Ford's production improvements from 1913 was revolutionary for the world car industry. In 1908, Ford's factories produced 18000 **Model Ts**, which sold at over US$1000. By 1923, this had increased a hundredfold with US production lines turning out 1.8 million Model Ts (popularly known as 'Tin Lizzies' or 'flivvers'). The price had dropped accordingly to under $300, and Ford had created a **mass market** for cars.

TIN LIZZIE More than 15 million Model Ts were sold in the USA in the 19 years between 1908 and 1927.

Computers 1

Core facts

◆ The earliest mechanical aid to calculating was probably the **abacus**, invented in China in about 2600 BC. Today, computers are used to order almost all aspects of modern life.
◆ The complexity, speed and power of computers have **more than doubled** every two years since the early 1980s ('Moore's Law').

◆ The greatest advances in computer technology have occurred because of improvements in the making and **miniaturisation** of components.
◆ Computers are now under development that will use optical, chemical or quantum (single-electron) activity rather than electricity, for **even higher speeds**.

TIMESCALE ❷

► **1642** French mathematician Blaise Pascal builds a calculating machine which performs addition and subtraction using gears.
► **1801** Frenchman Joseph-Marie Jacquard invents a punched-card system for programming looms.
► **1834** Englishman Charles Babbage conceives of an Analytical Engine, including many of the elements of modern computers, such as a memory to store numbers.
► **1890** Herman Hollerith devises a punch-card system to tabulate the US Census. In 1891 he forms the company that will become IBM.
► **1937** The British mathematician Alan Turing describes his Virtual Machine. His concepts of information storage and retrieval form the basis of future computers.
► **1943-5** US army builds ENIAC (Electronic Numerical Integrator and Computer) for calculating ballistic trajectories.
► **1944** IBM and Harvard University build the Mark 1 computer.
► **1945** US mathematician John Von Neumann conceives the computer as a general-

FORERUNNER Part of Babbage's Analytical Engine, still not completed when he died in 1871.

purpose digital device with stored programs.
► **1949** The Manchester Automatic Digital Machine ('MADAM') is the first computer with stored programs.
► **1971** Ted Hoff of Intel develops the microprocessor.
► **1977** Steve Jobs and Steve Wozniak launch the Apple II.
► **1981** IBM releases its Personal Computer.
► **1984** Apple releases its Macintosh.
► **1985** Microsoft releases its Windows operating system.
► **1998** Microsoft is the world's most valuable company, worth over US$260 billion.

WEIRD AND WONDERFUL ❸

Program errors are said to be caused by '**bugs**'. The term was coined by computer pioneer Grace Hopper in 1945 when the computer she was working on malfunctioned because a moth had got caught inside it.

Principles of computing ❹

Computers still work in the way envisaged by Turing, and share three features:
◆ an **input/output** device.
◆ **memory**.
◆ a central **processing** unit. Digital data is read from memory or input devices into the processing unit, where it is modified and relocated in memory. The results become output.

SPEAKERS PCs may have sound capabilities rivalling hi-fi equipment in quality.

MOUSE Allows the user to select data on the monitor.

KEYBOARD Encodes digitally the data keyed in by the user.

MONITOR Screen graphics – or GUI, graphical user interface software – allow the user to carry out commands.

Valves to chips ❺

All computer data exists as **binary** numbers (using 0 and 1) because electric current has only two states: off and on. Any operation requires fast switching between the two states, by many switches.

In **first-generation** electronic computers, this switching was done by vacuum tubes. These were unreliable, generated heat and required a lot of space. In 1947, William Shockley of Bell Laboratories in the USA invented the **transistor**, a much smaller switching device. **Second-generation** computers used these. **Third-generation** computers, from the late 1960s, had transistors and other components integrated in a single circuit board. **Fourth-generation** computers, from 1974, used **microprocessors**, with thousands of tiny transistors packed onto a single 'chip' of silicon. The chips in today's **fifth-generation** computers each contain millions of transistors.

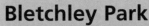
MINIATURISATION A microchip so small that an ant can hold it in its mouthparts.

Bletchley Park ❻

One of the most pressing concerns during the Second World War was to crack enemy secret codes. Decryption involved searching for patterns in seemingly random data – a mind-numbing task for hundreds of human cryptologists.

Led by **Alan Turing**, British scientists at Bletchley Park in Buckinghamshire designed the **Colossus** computer to take over this task. Operational from late 1943, Colossus contained 1500 vacuum tubes and was named for its physical size. It helped crack codes generated by the German **Enigma** machine and influenced the design of many later computers.

CODE-BREAKER Members of the Women's Royal Naval Service ('Wrens') operate the Colossus computer at Bletchley Park in England.

RAM CHIPS Microchips for storing programs and data in use. Data here is lost when the computer is switched off.

CENTRAL PROCESSING UNIT (CPU) Also known as the microprocessor. The microchip that processes data and coordinates input, output and storage devices.

FLOPPY, CD-ROM, DVD-ROM DRIVES Data-storage devices that use removable disks.

BIOS A chip that stores vital data even when the computer is switched off, including instructions for starting the computer.

HARD DISK DRIVE The main permanent store of programs and data. Data is stored as magnetic signals on metal disks. The data is 'read' from and 'written' to the fast-spinning disks by magnetic heads similar to those of a tape recorder.

MOTHERBOARD The main circuit board.

964 Silicon Valley ⭐

Silicon Valley is the San Francisco–San Jose corridor in California where hi-tech companies congregate. By 2002, 79 variants of the name were in use, including, in Britain, **Silicon Fen** (Cambridge) and **Silicon Glen** (from Glasgow to Edinburgh).

EXTERNAL STORAGE Supplementary storage for data.

EXTERNAL DRIVE With removable disks, as in a Zip drive.

CD-DVD DRIVE Input device for programs, games and movies.

PRINTER Often ink-jet, spraying a fine jet of fast-drying ink, or laser. ❼

SCANNER Photo-electric input device that encodes visual data such as pictures or printed text. ❽

Computers 2

PLAY STATION From their beginnings in the 1970s, video and computer games have become a major industry.

Computer games ❶

The earliest **video games** were played on a variety of purpose-built consoles – ancestors of the Sony, Sega, Nintendo and Microsoft game consoles of today. The **personal computer** became a major games platform only in the late 1990s, when its video capabilities began to rival those of the consoles. Games previewed at the E3 (Electronic Entertainment Expo) in 2002 suggest that a new generation of 'photo-realistic' games are now nearing reality. **Influential early games** include:

Game	Date	Developer	Notes
Pong	1972	Atari Corporation	Simple ball-and-paddle tennis simulation, released on many different consoles
Space Invaders	1978	Taito Corporation	Hugely successful console game requiring players to shoot waves of advancing aliens
Elite	1984	David Braben/ Ian Bell	Space exploration and trading game first released on the BBC Microcomputer. First genuine 3D game on home computer
Sim City	1989	Maxis	First of a long line of games in which players build their own cities or worlds, based on emergent, artificial intelligence interactions
Doom	1993	id Software (John Romero/ John Carmack)	Million-selling 'first-person shooter' game released partly on a shareware ('try before you pay') basis

Computer terminology ❷

BIOS Basic Input/Output System. Software on a separate chip that starts the computer's basic functions at bootup.
Bootup From 'pulling oneself up by one's bootstraps'. The computer's process of setting itself up when the power is turned on.
Byte From 'by eight'. A group of eight bits, or fundamental units of data.
Clock speed Also known as clock rate. The speed at which a processor works, usually expressed in megahertz (MHz).
GUI Graphical user interface. Software that uses visual images ('icons') on the screen to represent commands.
Modem Modulator-demodulator. A device that enables computer data to be sent by telephone line.
Network A system with two or more computers linked together to share data.
OS Operating system. Standard software that performs basic computer functions.
RAM Random-access memory. Memory used as a temporary store for data. Data in RAM can be overwritten, and unless 'saved' to disk, it is lost when the computer is turned off.
ROM Read-only memory. Permanent memory (often on a separate chip) that holds basic computer instructions. Data is retained when the computer is turned off.
VDU Visual display unit. A monitor – a television-like cathode-ray tube, or a liquid-crystal display (LCD) or plasma display.
Virus A hidden, self-replicating program that interferes with the behaviour of host computers.

⭐ **647**

PC or Mac?

In 1981, IBM made most aspects of the design of its **Personal Computer** available to competitors. The PC thus became an industry standard, squeezing other designs out of the market. Only **Apple**, with its innovative designs and hold on the creative industries, still supplies an alternative, the **Macintosh**.

Operating systems

	DOS	Unix	Windows	MacOS
Creator	Microsoft/IBM	AT&T	Microsoft	Apple
Date	1981(MS-DOS)	Late 1960s	1985 (Version 1.0)	1984
Current version	No longer available	Many variants	Windows XP	Mac OS-X
Use	PC	Mostly mainframe (very large) computers and network servers	PC	Macintosh
User interface	Command line (user types in a command on the keyboard)	Command line, but most versions provide a graphical user interface (GUI – see opposite)	Graphical user iinterface (GUI)	Graphical user iinterface (GUI)
Notes	Now subsumed into Windows	Freeware Linux version, created by Linus Torvalds (1992), used on some PCs	Originally a GUI for DOS, but now a full operating system. Overwhelmingly dominant in PCs	First successful GUI-based OS. Many find it easier to use than Windows

Talking to machines

All computer programs are written in **programming languages**. They range from low-level languages, like machine code (written entirely in binary numbers), through languages like Java that use terms nearer to English, to the highest level ones, like BASIC and COBOL.

The 'higher' a language, the more closely it relates to the needs of the programmer rather than the computer. But programs in higher languages still have to be 'translated' by special programs into machine language before the computer can execute them.

The future of computing

Computer technology has **advanced so quickly** that many relatively recent predictions are already reality. Several current trends point the way to the future.

◆ Computer chips will continue to shrink in size – ultimately to **atomic** or **sub-atomic** scale – and to find a role in ever more areas of life. Our entire domestic, commercial and industrial environments may soon be computer controlled.

◆ The distinction between humans and computers will begin to blur. **Neuro implants** will enhance human capabilities, while computers themselves will begin to include organic elements – memory may soon be 'grown', like human brain cells.

◆ Communications technology will ensure that all computers are ever more closely **integrated**. This will make **artificial intelligence** a reality – not the thinking robots of science fiction, but software that can learn, replicate and evolve, via the Internet.

FLIGHT CONTROL Human responses are not fast enough for pilots to control the Grumman X-29 fighter without the aid of its on-board computer.

ROBODOG Sony's Aibo robotic 'dog' can recognise 75 spoken commands – see also page 106.

COMPUTER TO WEAR The Poma computer, developed by Xybernaut, is small enough to fit into a pocket and has a head-mounted display unit.

The Internet revolution

Core facts ❶

◆ The Internet is a **network of computer networks**, constantly re-established and reconfigured while in use. It is collaborative and essentially unregulated.
◆ The Internet's fundamental technical idea is **packet switching**, devised in 1962 by Paul Baran. All files are communicated in small sections ('packets'), which are constantly switched from one route to another to maximise speed of transfer.
◆ Most individuals connect to the Internet using their computer's modem and a telephone line, but high-speed (**'high-bandwidth'**) connections are increasingly common.

TIMESCALE ❷

▶ **1969** US military and academic researchers establish the ARPAnet (Advanced Research Projects Agency Network).
▶ **1974** Telnet, the first commercial version of ARPAnet, is set up. ARPAnet founder Vint Cerf pioneers ways of linking separate networks into an Internet.
▶ **1984** The domain name system is established for identifying host computers on the Internet.
▶ **1991** The Internet is first opened to commercial traffic, and the first ISPs offer subscriber access.
▶ **1993** Marc Andreessen and Eric Bina create the Mosaic browser for web pages. Later, Netscape Navigator, developed from Mosaic, becomes the standard browser worldwide.
▶ **2000** The number of host computers reaches 95 million. Over 1 billion web pages are in existence.

CHILLY ACCESS
A fridge developed in South Korea also allows Internet access.

The World Wide Web ❹

The World Wide Web (**WWW**, **W3** or just **'Web'**) is a subset of the Internet. It consists of servers that use the Internet to supply over a billion web pages – files created in the HTML language (see 'Web jargon' below) and its successors. These files are viewed using browser software on client computers. Web pages may include graphics, animations, movie clips, sound files and programs.

★ **137**

Berners-Lee

British scientist Tim Berners-Lee was effectively the **inventor of the World Wide Web**. While working at CERN (Conseil Européen des Recherches Nucléaires) in Geneva in about 1990 he devised the URL, HTTP and HTML standards (see 'Web jargon' below) that made the Web possible.

WHERE THE WEB BEGAN Tim Berners-Lee's computer at the time he was pioneering the World Wide Web.

Web jargon ❸

Backbone A main link between the major networks that make up the Internet.
Email Electronic mail, sent over a network or the Internet.
HTML Hypertext Markup Language. The standard formatting language for web pages.
HTTP Hypertext Transfer Protocol. The procedures for communicating between servers and browsers on the Web.
Hypertext Web-page text that includes clickable links (hyperlinks) to other web pages.
ISP Internet Service Provider. A company that provides Internet access to subscribers.
JPEG Joint Photographic Experts Group. A compressed format, for transmitting graphics on the Web.
Server Any computer set up to supply files to other computers on a network.
Spam Slang for junk mail distributed via the Internet.
TCP/IP Transmission Control Protocol/Internet Protocol. A universal standard for Internet connection.
URL Uniform Resource Locator. The unique address of every web page.

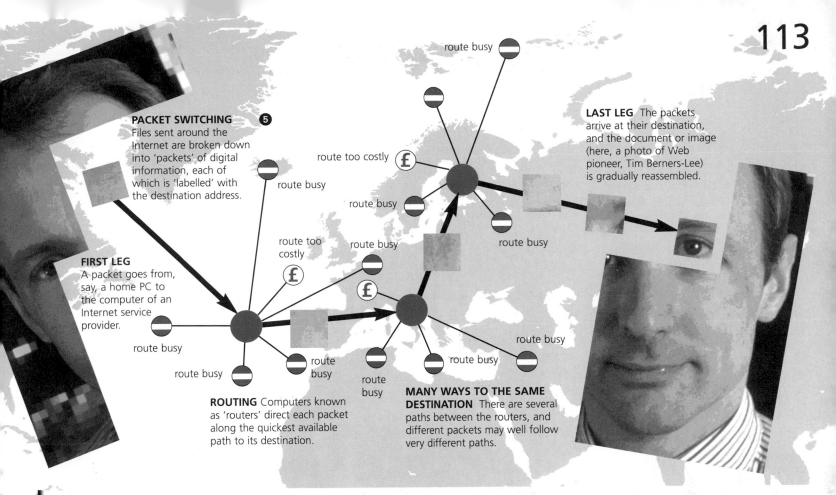

PACKET SWITCHING ❺
Files sent around the Internet are broken down into 'packets' of digital information, each of which is 'labelled' with the destination address.

route busy

route too costly

route busy

route busy

route busy

LAST LEG The packets arrive at their destination, and the document or image (here, a photo of Web pioneer, Tim Berners-Lee) is gradually reassembled.

FIRST LEG
A packet goes from, say, a home PC to the computer of an Internet service provider.

route too costly

route busy

route busy

route busy

route busy

route busy

route busy

route busy

route busy

route busy

ROUTING Computers known as 'routers' direct each packet along the quickest available path to its destination.

MANY WAYS TO THE SAME DESTINATION There are several paths between the routers, and different packets may well follow very different paths.

http://www.rd.com/common/nav

PROTOCOL
The first part of the URL indicates the appropriate file transfer method.

SERVER NAME The name of the particular server that holds the required file – 'www' indicates the domain's web server is required.

DOMAIN NAME
A unique identifier, which must be registered with Network Solutions, Inc.

DIRECTORY NAME The name of the directory that contains the required file on the server. Sub-directory names may follow, separated by slashes.

Domain name suffixes ❻

A domain is a group of **Internet-connected computers** that is administered as a unit. Usually, it is part of a single organisation or group of subscribers to a particular Internet service provider.

Domain name suffixes indicate the type of organisation involved. Some examples are:

- **.com** for a commercial organisation
- **.gov** for a government department or agency
- **.org** for a non-profit-making organisation – a charity, for example
- **.net** for an Internet administrator, Internet service provider or similar body
- **.edu** for an educational institution, typically a school
- **.ac** for an academic institution, typically a university
- **.co** for a commercial company (usually UK-based: **.co.uk**)

A further level of information is provided by geographical extensions to domain name suffixes, such as **.uk** (United Kingdom), **.ru** (Russia), **.to** (Tonga).

Search engines ❼

Search engines help users to find information on the Internet. Many sites compile huge indexes, using either programs called Web **crawlers** or **spiders**, or human editors. Others, like Yahoo, present users with hierarchical indexes that enable them to search the Web by subject.

Name	URL	Date set up	Notes
AltaVista	http://www.altavista.com	Dec 1995	Crawler-based engine
Google	http://www.google.com	Jun 1998	Largest crawler-based engine
HotBot	http://www.hotbot.com	May 1996	Set up by *Wired* magazine
LookSmart	http://www.looksmart.com	Oct 1996	Human-compiled directory
Yahoo	http://www.yahoo.com	1994	Human-compiled directory

Matter

Core facts ❶

◆ Matter is what all the solid, liquid and gaseous material in the Universe consists of.
◆ Matter is made up of **atoms**, **molecules** and **subatomic particles**.
◆ All matter has **inertia**, a tendency to stay put or move steadily. **Mass** – the amount of matter in an object – is a measure of inertia.
◆ Matter exerts a **gravitational pull** on other matter. This pull on an object is its weight.
◆ Most of the matter in the Universe is believed to be '**dark matter**' that cannot be detected because it does not emit any form of electromagnetic radiation (see page 86). Its presence is revealed by its gravitational effect.
◆ According to Einstein's Theory of Relativity, matter and energy are equivalent (see page 54).

What is an atom? ❷

Atoms are the **smallest particles of chemical elements** (see page 116) that have the properties of the element. Atoms can be split, but they then cease to be atoms of the original element.

Atoms are extremely small – typically one ten-millionth of a millimetre across – but they are made up of even smaller, subatomic particles. Most of an atom's mass is concentrated in its **nucleus**, around which orbit clouds of minute **electrons**. The number of electrons determine an atom's chemical character.

ELECTRON Tiny packets of energy with a negative electric charge, electrons move in shells (called orbitals) around the nucleus.

NUCLEUS Many times smaller than the whole atom, the nucleus contains almost all of an atom's mass. The nucleus consists of protons and neutrons.

QUARKS Protons and neutrons each consist of a packet of three smaller particles called 'up' and 'down' quarks. These are held together by gluons.

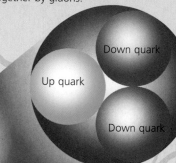

Down quark
Up quark
Down quark
Neutron
Proton

WATER The formula H_2O indicates that a water molecule consists of one oxygen (O) atom linked to two hydrogen (H) atoms.

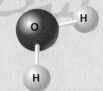

O H H

What is a molecule? ❹

Most substances consist of more than one atom. Such **groups of atoms**, linked together by chemical bonds, are called **molecules**.

Some molecules contain only one type of atom; for example, an oxygen molecule contains two oxygen atoms. Most molecules contain more than one type of atom.

The formula used to describe a chemical shows, in abbreviated form, how many of which atoms one molecule contains.

SULPHURIC ACID The formula H_2SO_4 shows that a molecule of sulphuric acid contains two hydrogen (H) atoms, one sulphur (S) atom and four oxygen (O) atoms.

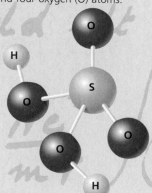

O H S O O H

States of matter ❺

Throughout the Universe, matter exists in three main forms, or states: **solid**, **liquid** and **gas**. The molecules in any substance are arranged differently and interact in different ways in the three states.

◆ **Solid** The molecules are arranged rigidly in three dimensions. They can vibrate but cannot move their positions.

◆ **Liquid** The molecules cling together loosely, but can move and slip past each other quite freely, which allows liquids to flow.

◆ **Gas** The molecules in a gas are more widely spaced than in solids and liquid states, and move at high speeds.

Solids have a **fixed shape and size** and are generally more or less hard. Liquids and gases can flow, and are known as fluids. **Liquids** have a **fixed volume**, but gases do not. Liquids fill a container to a particular level depending on the volume of liquid. A gas let into a rigid empty container will expand to fill the container however big it is.

Most substances turn from solid to liquid and from liquid to gas at specific temperatures – known as **melting** and **boiling** points. Because molecules in a liquid (and a gas) move around more than those in a solid, they carry more energy (see page 60). This energy has to be supplied (in the form of heat) in melting or boiling. Known as **latent heat**, this heat does not increase temperature, so melting ice and the water it forms are both 0°C.

WARMING UP Every substance has its own melting and boiling points at normal atmospheric pressure. These points vary immensely between substances, as the graph shows.

■ Boiling point
□ Melting point

Temperature °C

				2750
2212				
		1535		
	962			
357				
	100			
−39	0			
−272 −269				
Helium	Mercury	Water	Silver	Iron

Melting and boiling points are affected by pressure, and at some pressures certain solids turn directly into gases without forming a liquid; this is called **sublimation**. For example, at normal atmospheric pressure solid carbon dioxide ('dry ice') turns directly to a gas; if it is heavily compressed, it forms a liquid, even at room temperature.

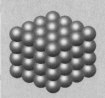

SOLID In a solid, the atoms are tightly packed, often forming a regular structure, and unable to move freely.

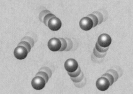

LIQUID Atoms or molecules in a liquid are held together quite tightly, but can slip past one another. This is why liquids flow.

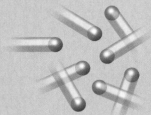

GAS In a gas, the atoms or molecules are not held together at all. Unlike solids or liquids, gases expand to fill an empty space.

Splitting the atom ❻

For centuries, scientists believed that atoms were the smallest particles and could not be split. But in 1919, a New Zealander, Ernest Rutherford, **first artificially disintegrated** (or split) an atom. In 1932 John Cockroft and Ernest Walton first split lithium atoms by bombarding them with protons.

In 1938, Lise Meitner, Otto Hahn and Fritz Strassmann discovered the splitting of uranium nuclei by neutrons. This led to the development of **atomic weapons** and **nuclear power**.

NUCLEAR DEVICE Ernest Rutherford with his atomic disintegration apparatus.

Mass, volume and density ❼

In general, matter expands as it gets hotter. Gases consist of small molecules flying around in a space, so their density (mass per unit volume) depends on the mass of the molecules and the size of the container. In solids and liquids molecular mass is a factor, but a substance's density mainly depends on how tightly the molecules are packed. Most solids are more tightly packed, so denser, than their liquid form. **Water** is an exception. As it cools below 4°C it expands slightly. Ice is less dense still, which is why ice floats.

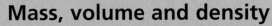

★ 744

Mimicking creation

In a vast underground circular tunnel (right) at CERN (Conseil Européen des Recherches Nucléaires), in Geneva, subatomic particles are accelerated almost to the speed of light, then made to collide. The reactions produce new particles and mimic conditions from the Universe's early history, helping to reveal the nature of matter.

116

QUESTION NUMBER

The numbers or star following the answers refer to information boxes on the right.

ANSWERS

- **66** B: Stibnite (it is a mineral, antimony sulphide) ❷
- **73** Neon ❷
- **130** Helium ❷
- **262** True (it is graphite, one of the forms of carbon) ❷
- **301** Hydrogen ❷❸❺
- **310** Isotopes ❶
- **382** Magnesium ❷
- **386** Nitrogen ❷❺
- **429** Barium ❷
- **461** Calcium ❷
- **464** Iron ❷
- **465** Mercury ❷
- **467** Copper (which was plentiful in Cyprus) ❷❸
- **468** Zinc ❷
- ★ **512** A: Dmitri Mendeleyev ★
- **609** The Palladium ❷
- **648** An atom of tin ❷
- **669** Nickel and cadmium ❷
- **808** Phosphorus ❷
- **847** Metallic elements ❷

The elements

Core facts ❶

◆ Elements are the simplest and most basic **chemical substances**; they cannot be split down chemically into anything simpler.
◆ A total of **112 elements** are known, but only about 90 occur in nature. The rest have been produced by nuclear reaction.
◆ All atoms in a sample of a **pure element** are indistinguishable chemically.

◆ All atoms of an element have the same **atomic number** – the number of protons or electrons in the atom (see page 114). The number of neutrons can vary. Atoms of an element with extra neutrons are called **isotopes**.
◆ The **periodic table**, in which the elements are laid out according to their chemical properties, was constructed by **Mendeleyev**.

metals
non-metals
transition elements
metalloids

The periodic table ❷

The table lists the 112 elements, arranged in order of **atomic number**. Each element's symbol is given, together with its **relative atomic mass** – the weight of one atom of an element relative to the weight of an atom of any other element. Where an element has no stable isotopes, the mass number of the longest-lived isotope is given in brackets.

Each horizontal row represents a **period**. Reading from left to right across a period, the number of electrons in the elements increases. Also, the elements change from being metallic to non-metallic.

Eight vertical columns or **groups**, numbered I-VIII, are arranged down each side of the table. Atoms of the elements within each group have the same number of electrons (which govern chemical reactions) around the nucleus so react in similar ways.

The **transition elements** are metals with similar outer orbital shells; extra electrons are added to lower shells. These elements are good conductors of heat and electricity. **Metalloids** contain some properties of metals and some of mon-metals.

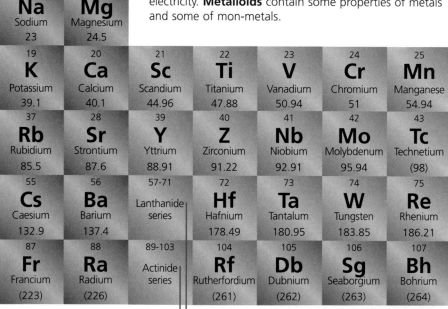

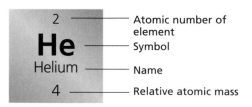

2 —— Atomic number of element
He —— Symbol
Helium —— Name
4 —— Relative atomic mass

Elementary records ❸

- ◆ Densest metal: osmium
- ◆ Densest non-metal: iodine
- ◆ Least dense metal: lithium
- ◆ Least dense non-metal solid: hydrogen
- ◆ Highest melting point: tungsten (3410°C)
- ◆ Lowest melting point: helium (−272.2°C)
- ◆ Best conductors: (1) silver; (2) copper
- ◆ Most corrosion-resistant: iridium
- ◆ Most malleable: gold
- ◆ Hardest: diamond (a form of carbon)
- ◆ Most common in Universe: hydrogen
- ◆ Most common solid element in Earth's crust: silicon (27.7 per cent)

Made by man ❹

More than 20 new elements, and dozens of isotopes of natural elements, have been made artificially. Many are by-products of nuclear reactions, and most are highly radioactive. A few have been made by accelerating elementary particles (see page 114) into the nuclei of heavy atoms so that they fuse. For example, the heaviest element properly named, meitnerium (Mt), was made by fusing iron and bismuth nuclei.

Elements of life ❺

The body's composition varies with how fat it is. (Fat consists mainly of carbon, hydrogen and oxygen atoms.) The following percentages are those of an average adult.

- ■ Oxygen: 65 per cent
- ■ Carbon: 18 per cent
- ■ Hydrogen: 10 per cent
- ■ Nitrogen: 3 per cent
- ■ Calcium: 1.5 per cent
- ■ Others: 2.5 per cent

★ 512

The man behind it all

Russian chemist **Dmitri Mendeleyev** (1834-1907) published his first version of the periodic table in 1869. He left gaps for then-unknown elements, which were found later. Mendelevium (Md), discovered in 1955, is named after him.

					VIII
					2 **He** Helium 4

III	IV	V	VI	VII	
5 **B** Boron 10.81	6 **C** Carbon 12.0	7 **N** Nitrogen 14.01	8 **O** Oxygen 16	9 **F** Fluorine 19	10 **Ne** Neon 20.18
13 **Al** Aluminium 26.98	14 **Si** Silicon 28.09	15 **P** Phosphorus 30.97	16 **S** Sulphur 32.06	17 **Cl** Chlorine 35.45	18 **Ar** Argon 39.94

26 **Fe** Iron 55.85	27 **Co** Cobalt 58.93	28 **Ni** Nickel 58.69	29 **Cu** Copper 63.55	30 **Zn** Zinc 65.38	31 **Ga** Gallium 69.72	32 **Ge** Germanium 72.6	33 **As** Arsenic 74.92	34 **Se** Selenium 78.96	35 **Br** Bromine 79.9	36 **Kr** Krypton 83.8
44 **Ru** Ruthenium 101.07	45 **Rh** Rhodium 102.91	46 **Pd** Palladium 106.42	47 **Ag** Silver 107.87	48 **Cd** Cadmium 112.41	49 **In** Indium 114.82	50 **Sn** Tin 118.69	51 **Sb** Antimony 121.7	52 **Te** Tellurium 127.6	53 **I** Iodine 126.9	54 **Xe** Xenon 131.29
76 **Os** Osmium 190.2	77 **Ir** Iridium 192.22	78 **Pt** Platinum 195.08	79 **Au** Gold 196.97	80 **Hg** Mercury 200.59	81 **Tl** Thallium 204.38	82 **Pb** Lead 207.2	83 **Bi** Bismuth 208.98	84 **Po** Polonium (209)	85 **At** Astatine (210)	86 **Rn** Radon (222)
108 **Hs** Hassium (265)	109 **Mt** Meitnerium (266)	110 **Uun** Ununnilium (369)	111 **Uuu** Unununium (266)	112 **Uub** Ununbium (277.15)						

61 **Pm** Promethium (145)	62 **Sm** Samarium 150.36	63 **Eu** Europium 151.96	64 **Gd** Gadolinium 157	65 **Tb** Terbium 158.93	66 **Dy** Dysprosium 162.50	67 **Ho** Holmium 164.93	68 **Er** Erbium 167.26	69 **Tm** Thulium 168.93	70 **Yb** Ytterbium 173.04	71 **Lu** Lutetium 174.97
93 **Np** Neptunium (237.05)	94 **Pu** Plutonium (244)	95 **Am** Americium (243)	96 **Cm** Curium (247)	97 **Bk** Berkelium (247)	98 **Cf** Californium (251)	99 **Es** Einsteinium (252)	100 **Fm** Fermium (257)	101 **Md** Mendelevium (258)	102 **No** Nobelium (259)	103 **Lr** Lawrencium (260)

Compounds and reactions

Core facts ❶

◆ A **molecule** is the smallest unit of a chemical substance that can take part in a reaction.
◆ Although a molecule may consist of a single atom, most have two or more atoms (see page 114). The biggest contain millions of atoms.
◆ If a substance contains more than one type of atom, it is known as a **compound**.
◆ The atoms in a molecule are held together by **bonds** of various types, which can be broken or changed in chemical reactions.
◆ Atoms in a molecule cannot be separated by simple physical means such as filtering.

Types of compound ❷

Compounds are traditionally classified as **organic** (compounds of carbon) and **inorganic** (all other compounds). But they may also be classified according to where they are found, their characteristic atoms, or the type of bonds that link their atoms. The divisions are largely a question of convenience. In 1828, the German chemist **Friedrich Wöhler** disproved the absolute division between organic and inorganic chemistry when he made 'organic' urea from 'inorganic' ammonium cyanate; they have the same atoms arranged differently – $CO(NH_2)_2$ and NH_4CNO.

	Examples	Defining characteristics
Inorganic compounds	Water (H_2O); salt (sodium chloride, NaCl); chalk (calcium carbonate, $CaCO_2$)	Traditionally regarded as 'non-living', these include all substances except most compounds of carbon, although simple carbon compounds such as carbonates are regarded as inorganic. Atoms can be linked by ionic, covalent or metallic bonds.
Organic compounds	Methane (CH_4); ethene (C_2H_2); ethanol (C_2H_5OH); vinegar (acetic acid, C_2H_5COOH)	Traditionally thought to be components and products of living things, these all contain carbon atoms – often linked in chains or rings – with hydrogen and, usually, other atoms attached. They are mainly linked by covalent bonds, but ionic bonds occur in organic acids (such as acetic acid) and their compounds.

Chemical bonds ❸

Various types of bonds hold together the individual atoms in compound molecules. They all depend on the behaviour of the electrons orbiting around the nucleus (see page 114).

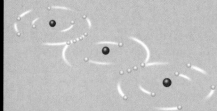

COVALENT BONDS Two adjacent atoms share pairs of electrons (one from each atom) that, in effect, orbit around both atoms, bonding them strongly.

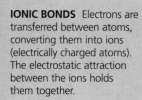

IONIC BONDS Electrons are transferred between atoms, converting them into ions (electrically charged atoms). The electrostatic attraction between the ions holds them together.

METALLIC BONDS Atoms lose their outer electrons, which form a 'sea' of electrons around the metal ions. The electrons move and act like electrostatic 'glue'.

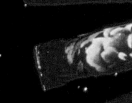
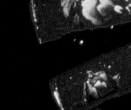

Formulae and equations ❹

A **chemical formula** is shorthand for the types and numbers of atoms in a molecule. It also shows how they are arranged. For example, the formula for ammonium nitrate is NH_4NO_3. It has two nitrogen (N) atoms in each molecule, but one is part of an ammonium group (NH_4) while the other is part of the nitrate group (NO_3).

A **chemical equation** shows the changes taking place in a reaction. By using formulae, and making sure that the equation is 'balanced' (the same number of each type of atom are on each side) you can see how many molecules take part and exactly how they react. For example, copper dissolves in nitric acid to make copper nitrate, nitric oxide and water. The equation (below) shows that a copper atom reacts with two molecules of sulphuric acid to make one molecule of copper sulphate, one of sulphur dioxide and two of water.

$$Cu + 2H_2SO_4 \rightarrow CuSO_4 + SO_2 + 2H_2O$$

Acids and bases ❺

Acids are substances that make hydrogen ions (H^+, hydrogen atoms with a positive electrical charge) in water. They taste sour and attack many metals to form salts. **Bases** are their chemical 'opposite', reacting with hydrogen ions to form neutral substances (neither acids nor bases).

Some bases dissolve in water to form negatively charged hydroxyl ions (OH^-): they are called **alkalis**. They taste bitter and feel slippery, turning skin oils into a soap-like substance. The strength of an acid or alkali is measured on the **pH** scale (right); the higher the pH number, the more alkaline it is, the lower the more acidic. A pH value of 7 is neutral.

Hydrochloric, nitric, sulphuric acid	0
	1
Stomach juices	
Lemon juice	2
Vinegar, sour apple	3
Acid rain	4
Typical rainwater	5
Milk, urine, saliva	6
Distilled water	7
Blood	
	8
Sea water	
Baking soda	9
Antacid tablet	10
Ammonia	11
Soapy water	12
Oven cleaner	13
Caustic soda	14

Acidic — Alkaline

⭐ 193

Going with a bang!

Gunpowder was invented in China in about the 7th century, but not seen in the West until 600 years later. It contains powdered potassium nitrate (nitre, or saltpetre), sulphur (brimstone) and carbon (charcoal). Triggered by a spark, the potassium nitrate oxidises the sulphur and carbon very rapidly. This produces the gases nitrogen, carbon dioxide and carbon monoxide; the heat released makes the gases expand explosively.

Reactions all around us ❻

◆ **Burning** Natural gas is mainly methane (CH_4), which burns with oxygen in the air to make carbon dioxide and water vapour: $CH_4 + 2O_2 \rightarrow CO_2 + 2H_2O$.

◆ **Corrosion** Some metals react with oxygen to form a metal oxide – a process called oxidation. Rust is an oxide of iron, but rusting is a complicated process, needing water and carbon dioxide as well as oxygen. The tarnishing of silver is simpler. Silver reacts with hydrogen sulphide (H_2S) plus oxygen in the air to make black silver sulphide (Ag_2S): $4Ag + 2H_2S + O_2 \rightarrow 2Ag_2S + 2H_2O$.

◆ **Baking** Baking powder contains sodium bicarbonate ($NaHCO_3$) plus a weak acid, such as tartaric acid. In cooking, the bicarbonate reacts with the acid to release carbon dioxide (CO_2) gas, which makes the cake rise.

◆ **Bleaching** Both hydrogen peroxide (H_2O_2) and chlorine bleach release 'active' oxygen – single oxygen atoms – which then oxidise stains and dyes to colourless substances.

Types of reaction ❼

Industrial chemistry often involves splitting compounds into simpler parts (**decomposition**), or building them into more complex substances (**combination** or synthesis), as when hydrogen and nitrogen are combined to make ammonia.

In some processes, atoms of one element displace those of another – when oxide ores are smelted, carbon displaces metal atoms to give carbon dioxide and the free metal. When chemical bonds are broken, energy is released (as in burning) but far less than when nuclear bonds are broken (see pages 68 and 114).

Radioactivity

Core facts ❶

◆ Radioactivity is the **emission of radiation** – invisible 'rays' – by natural or artificial elements.
◆ Naturally radioactive elements or isotopes of elements (see page 116) are mostly those with **high atomic weights**, such as uranium and radium. Other elements can be made radioactive by bombarding them with radiation.

◆ Nuclear reactions include **fission** (splitting), **fusion** (joining) and **transmutation** (converting into a different nucleus).
◆ Radiation consists of particles and electromagnetic waves. It is **dangerous** because it makes charged ions that damage living cells and cause mutations.

Fission and fusion ❷

Fission splits an atom's nucleus into smaller fragments. In **fusion**, small nuclei are driven together so violently – usually by the heat of a fission explosion – that they join together. This releases even more energy than fission, but fewer radioactive by-products.

FUSION A tritium and a deuterium nucleus collide (in the centre) to form a larger helium nucleus. A neutron is lost in the process, releasing energy.

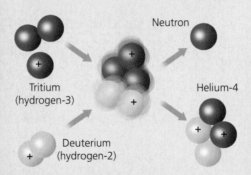

Neutron

Tritium (hydrogen-3)

Helium-4

Deuterium (hydrogen-2)

Alpha, beta, gamma ❸

Unstable radioactive nuclei emit subatomic particles (see page 114) or radiation. There are three types. **Alpha** particles consist of two protons and two neutrons (the same as the nucleus of a helium atom). **Beta** particles are fast-moving electrons. **Gamma** rays are a form of electromagnetic radiation (see page 87), like high-energy X-rays.

ALPHA Stopped by a sheet of paper.

BETA Stopped by 5 mm thick aluminium.

GAMMA Some can penetrate 4 cm thick lead.

Who's who in atomic physics ❹

◆ **Luis Alvarez** (1911-88) Developed the bubble chamber for observing short-lived particles formed during nuclear reactions.
◆ **Henri Becquerel** (1852-1908) In 1896 discovered radioactivity by accidentally 'fogging' a photographic plate with uranium.
◆ **Niels Bohr** (1885-1962) Developed the quantum theory of atomic structure.
◆ **Sir James Chadwick** (1891-1974) Discovered the neutron and helped to develop nuclear weapons.

◆ **Enrico Fermi** (1901-54) Discovered the transmutation of elements by neutrons. Built first nuclear reactor.

EXPLAINING ATOMS Danish physicist Niels Bohr won the Nobel prize for physics in 1922.

◆ **Hans Geiger** (1882-1945) Invented the Geiger counter, used to detect alpha, beta and gamma rays.
◆ **Otto Hahn** (1879-1968) With Fritz Strassmann (1902-80) and Lise Meitner (1878-1968) discovered the nuclear fission of uranium.
◆ **Ernest Rutherford** (1871-1937) Founder of nuclear physics. Discovered alpha and beta radiation, and atomic nucleus. Developed nuclear model of the atom.
◆ **Sir Joseph John Thomson** (1856-1940) Founder of particle physics. Discovered the electron.

Radioisotope dating ⑤

The rate of decay of radioisotopes – their half-life – can show how old objects are. Radiocarbon dating uses the decay of **carbon-14** (C-14). This isotope is constantly replenished in the atmosphere by cosmic rays bombarding nitrogen atoms, so the air (and living things) contain a constant amount. But as soon as something dies, its C-14 content starts to fall as C-14 decays to nitrogen-14, with a half-life of 5730 years. Measuring the C-14 content can show the age of an object since death, up to about 50000 years.

RADIOACTIVE DIAGNOSIS
Scintigrams, made by injecting the bloodstream with minute quantities of radioactive 'tracer' material, are used in the diagnosis of bone cancer.

⑥ WEIRD AND WONDERFUL
High-energy particles moving through a transparent material such as water make it glow. Known as **Cerenkov radiation**, it is the light equivalent of a sonic boom, caused by the particles travelling faster than light in water.

★ 403

End of the line

The main naturally occurring long-lived radioisotopes will all eventually decay to form **stable isotopes of lead**. Uranium-238 and radium-226 eventually become lead-206. Uranium-235 (and man-made plutonium-239) turn into lead-207, and thorium-232 ends up as lead-208. The long-lived man-made neptunium-237 and uranium-233 become bismuth-209.

Marie Curie ⑦

Born Marya Sklodowska in 1867, Marie married **Pierre Curie** (1859-1906) and shared a 1903 **Nobel prize** with him and **Henri Becquerel** for their work on radioactivity. She won a second prize in 1911, largely for isolating radium, but died of leukaemia in 1934 resulting from her life-long work with radioactive materials. Her daughter **Irène Joliot-Curie** (1897-1956) shared a 1935 Nobel prize with her husband **Jean-Frédéric Joliot-Curie** (1900-58) for work on artificial radioactivity, and also died of leukaemia.

LAB WORK
Marie Curie at work in her laboratory in Paris around 1908. She was the first person to win a Nobel prize twice, in 1903 and 1911.

Radioactive decay and half-life ⑧

When a radioactive isotope (**radioisotope**) gives off an alpha or beta particle (and often gamma rays), it changes, or decays, into an isotope of another element. **Half-life** is a measure of how fast isotopes decay. The decay of any one atom is random, but in a certain time – the half-life – half of the atoms in a piece of material decay. In the same time again, half the remainder decay, and so on. Each radioisotope decays in a particular way, with a half-life varying from fractions of a second to billions of years. Eventually, a **stable element** such as lead forms.

DECAY CHAIN
Thorium-232 becomes ten different isotopes, each with a different half-life, before stabilising as lead-208.

Thorium-232	Radium-228	Actinium-228	Thorium-228	Radium-224	Radon-220	Polonium-216	Lead-212	Bismuth-212	Polonium-212	Lead-208
14×10^9 years	5.8 years	6.1 years	1.9 years	3.6 days	55 seconds	0.15 seconds	11 hours	61 minutes	300×10^{-9} seconds	stable

Mining and oil extraction

Core facts ❶

◆ Mining and mineral extraction are among mankind's **oldest industries**, and our industrial civilisation still depends on them.
◆ Eras of pre-history and early history are defined by the **mineral basis** of technology: **stone**, **bronze** (at first copper and arsenic, later copper and tin) and **iron**.
◆ Mined materials are classified as **precious metals** and **minerals**; **ferrous metals** (iron and related metals); **non-ferrous metals**; and **non-metals**.
◆ Over 2500 minerals have been found so far.

Some uses of minerals ❷

PRECIOUS
Diamonds
Gems; cutting and grinding tools (often now synthetic)
Gold
Bullion; jewellery; electrical contacts
Platinum
(also iridium, palladium, osmium, rhodium and ruthenium) Used as catalysts in industrial processes and car exhausts
Silver
Photography; decorative uses; coins; electric circuits; batteries

FERROUS
Chromium
Hardening and plating steel
Cobalt
(and manganese, molybdenum, nickel, etc) Steel; plating; catalysts
Iron
Alloyed with small amounts of carbon, silicon and other metals to make a range of steels
Tungsten
Light bulbs; temperature-resistant uses

NON-FERROUS
Aluminium
Alloys used in ships, aircraft, cables, cans and kitchen utensils
Copper
Cables; brass and bronze
Lead
Batteries; bearings; radiation shield
Magnesium
Alloys with zinc and aluminium
Tin
Plating; alloys
Titanium
Spacecraft; high-speed aircraft; pigment (oxide)

Uranium
Nuclear fuel; ballast; weapons
Zinc
Plating; alloys

NON-METALS
Coal
Chemicals; smelting; fuel
Fluorspar
Steel industry; plastics
Phosphates
Fertiliser
Potash
Fertiliser; detergents; glass
Sulphur
Sulphuric acid

Deep mining ❸

◆ **Underground mining**
Methods depend on the **ore body**. A vertical or horizontal **shaft** is dug into the ground, with levels or drifts branching off. Ore is extracted using the **room-and-pillar** method (pillars left to support the roof); **longwall mining** (in strips); or **caving** (undercutting so the ore body collapses).
◆ **Underwater mining**
Mineral-bearing sand, gravel or mud is dredged by a scoop or chain of buckets.
◆ **Fluid methods** 'Native' sulphur and salt are extracted from deep underground by **pumping water** (superheated to melt sulphur) down a pipe and back up through a second pipe. Salt and magnesium minerals are extracted from evaporated seawater.

Surface methods ❹

◆ **Open-cast mining** Explosives and heavy machinery are used to remove 'overburden' (surface rock above the target ore), and to dig ore from thick seams close to the Earth's surface.
◆ **Strip mining** Shallow minerals are dug out in successive **parallel strips**. As each strip is removed, overburden is used to fill in the previous hole.
◆ **Quarrying** A similar process to open-cast mining, quarrying is used to extract minerals, where there is little or no overburden.

DIAMOND MINE
The Argyle mine in Western Australia produces more diamonds than any other mine in the world – 5.3 tonnes in 2000, mostly for industrial uses, rather than jewellery. It is an open-pit mine. The rock is drilled and blasted, then taken away in trucks to a processing plant, where the diamonds are sorted out from the waste material.

Major oilfields

Oil facts ❺

◆ Two-thirds of known oil reserves are in the **Middle East**, a quarter in Saudi Arabia alone.
◆ **Offshore oil** became increasingly important after the 1940s, as land-based stocks dwindled; up to a third now comes from offshore fields.
◆ The longest oil **pipeline** takes oil from the Urals 4000 km (2500 miles) to Germany.

Top oil producers
(million barrels/day)

Saudi Arabia*	8.22
Russia	6.76
United States	5.82
Iran*	3.71
China	3.28
Norway	3.16
Mexico	3.07
Venezuela*	2.91
Iraq*	2.50
UAE*	2.31
United Kingdom	2.28
Nigeria*	2.20
Kuwait*	2.08

* members of OPEC

TIMESCALE ❻

▶ **c.8000 BC** Copper first used in Mesopotamia.
▶ **c.6000 BC** Flint mined in pits and tunnels.
▶ **c.1400 BC** Systematic iron smelting in Asia Minor: start of Iron Age.
▶ **370 BC** First known use of coal as fuel, in China.
▶ **16th century AD** Native Americans use oil as fuel.
▶ **1821** First natural gas well, in New York state.
▶ **1859** First oil well drilled, in Pennsylvania.
▶ **1908** First big Middle East oilfield found in Iran.
▶ **1913** First oil cracking process invented.
▶ **1947** First offshore oil platform (Gulf of Mexico).
▶ **1948** Al Ghawar oilfield (world's biggest) found in Saudi Arabia.
▶ **1969** Oil first found in North Sea.

⭐ 560

Big rock

The **Cullinan diamond**, the world's biggest ever discovered, was found in South Africa in 1905. From it was cut the 530.2 carat Great Star of Africa, the 317.4 carat Lesser Star of Africa and 104 other almost flawless cut gems. They all form part of the British Crown Jewels. (1 carat = 200 milligrams.)

OIL RIG A platform in the North Sea.

Oiling the wheels of progress ❼

Crude oil is a complex and variable mixture of hydrocarbons, known as **fractions**. Each fraction vaporises (boils) and condenses at a different temperature. In an **oil refinery**, oil is boiled and then cooled as it rises through a distillation tower.

The various fractions condense at different heights, and are piped away for use. In **catalytic crackers**, less valuable heavy fractions are broken down into more valuable products such as aviation fuel and petrol. The final residue is **bitumen** (tar).

Gases	**Petrol**	**Kerosene**	**Diesel oil**	**Heating oil**	**Lubricating oil**	**Fuel oil**	**Bitumen**
(eg methane; ethylene; butane) Fuel; petrochemical raw materials	Fuel for cars, motorcycles and piston-engined aircraft	(paraffin) Jet aircraft fuel; heating and lighting; solvent	(gasoil) Fuel for lorries, tractors, some cars and smaller ships	Fuel for boilers in some houses and other buildings	Used to reduce friction in engines and other machinery	(bunker oil) Heavy-grade oil used as fuel in many larger ships	Semi-solid fraction used for road surfacing and waterproofing

Man-made materials

Core facts ❶

◆ People have developed a wide range of **artificial materials** over the centuries. These have been used to create everything from clothing and weapons to buildings and art.
◆ Some of the first new materials created were **metals**. Bronze – an alloy of copper and tin – was first made more than 5000 years ago. New alloys are constantly being developed.

The latest include metallic glasses; crystalline alloys and metals bonded with ceramics.
◆ Colours from natural sources have long been used to make **inks**, **dyes** and **paints**. Many pigments are now produced artificially.
◆ As well as metals, such as steel, numerous man-made structural materials have been created, including **glass** and **plastics**.

Paper ❷

The Egyptians were making rough paper from the pith of the **papyrus** reed as early as 3500 BC. Papyrus was lost to Europe after the fall of the Roman Empire, and medieval literacy depended on **parchment** – made from dried and cleaned animal skins, which were difficult to prepare.

True paper – made from vegetable fibres or rags pulped and mixed with water, then dried in flat sheets – was invented in China in the 2nd century BC. The technique reached Europe via the Middle East in the 8th century AD. Paper-making was automated in the 19th century. **Wood pulp** became the main ingredient in about 1840.

★ 139

The origin of mauve

In 1853, the British chemist **William Perkin** was attempting to synthesise quinine (used to combat the tropical disease malaria). Combining a coal-tar amine with a salt of the natural base aniline, he found that the resulting black sludge produced a strong purple colour. He had created the **first synthetic** dye. In 1856, Perkin set up a factory to manufacture the new dye, 'mauve', revolutionising the dyestuff industry.

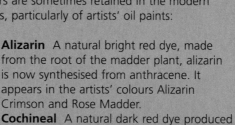

ONE MAN'S WORK Bottles of alizarin and mauve dye next to fibres dyed with mauve. William Perkin, who created mauve, also played a major part in discovering alizarin.

Creating colour ❸

Both dyes for fabrics and the pigments used to colour paints began as natural substances, but almost all have now been replaced by synthetic alternatives – many derived chemically from coal-tar. The pre-industrial, natural sources for artificial colours are sometimes retained in the modern names, particularly of artists' oil paints:

Alizarin A natural bright red dye, made from the root of the madder plant, alizarin is now synthesised from anthracene. It appears in the artists' colours Alizarin Crimson and Rose Madder.

Cochineal A natural dark red dye produced from carminic acid from the crushed bodies of female scale insects (*Coccus cacti*). Cochineal is used in Carmine Red, and the (now synthetic) food colouring also called Cochineal.

Indigo A natural blue or purple dye originally made from the leaves of the indigo plant. Indigo is now synthesised using aniline; the colour used for dyeing blue jeans.

Ochre Ochre included a wide range of yellows, browns and reds made from a mixture of iron oxide and clay, now synthesised (as 'Azo dyes') from aromatic amines. Ochre-based paints include Yellow Ochre, Raw Sienna, Burnt Umber.

Prussian blue The term refers to a range of deep blue pigments made from iron cyanides. It was first developed in 1704.

Glass

④

Glass is made from silica sand, fused by heating, with chemicals (such as lead, potassium, soda) added for particular qualities. It was **discovered** around 3000 BC, in the Middle East.

Glass-blowing was developed in Syria in the 1st century BC, and was used extensively in the Roman Empire. Glass became rare after Rome's fall, but stained-glass enlivened the windows of medieval cathedrals. Crystal was invented in Venice in the 15th century.

Plate glass was produced from the 17th century, by pouring molten glass onto an iron table. It then required polishing. Its successor, float glass, was invented by British glass-maker Alastair Pilkington in 1952. To create float glass, molten glass is poured onto molten tin, which gives it a completely flat surface.

PRECISION JOB A glassblower applies the finishing touches to a gigantic flask.

Fibreglass

⑤

Fibreglass – fibre made by drawing molten glass into filaments – was first demonstrated by US glassmaker **Edward Libbey** in 1893, but not manufactured until the 1930s. Fireproof, corrosion resistant and an excellent insulator, fibreglass is woven into industrial fabric or used for domestic thermal insulation; it is also bonded with resin to produce moulded shapes.

Plastic mouldings are also reinforced with carbon fibre. Researchers at Britain's Royal Aircraft Establishment discovered in 1963 that heat treatment of an acrylonitrile plastic produced fibres of carbon. Bonded with resin, these make a plastic as strong as high-tensile steel and one-quarter the weight. It is widely used in the aircraft industry.

MAGNIFIED IMAGE
Much modern fibreglass is a composite of glass fibres in polyester plastic.

WEIRD AND WONDERFUL

⑥

Tyre in Lebanon was famous for the dye known as **Tyrean purple**. Reserved for the robes of the Roman emperor, it was made by crushing the molluscs *Pupura* and *Murex*, whose rotting remains made the city stink.

BACKGROUND IMAGE: The US Federal Judiciary building in Washington DC was constructed with Pilkington planar glass. Planar glass enables vertical curtain walls to be built with a minimum of structural support.

Rubber

⑦

Natural rubber, the latex of several South American tree species, was known to the Aztecs in the 6th century. It was named 'rubber' by British scientist **Joseph Priestley** in 1770. Its tendency to soften when warmed was eliminated by vulcanisation (heating with sulphur), developed by US inventor **Charles Goodyear** in 1839, and it became a major industrial product because of its elastic and waterproof qualities.

Synthetic rubber, mostly SBR (styrene-butadiene rubber) was developed in the 1930s; cheaper and more durable, it is preferred for products like tyres.

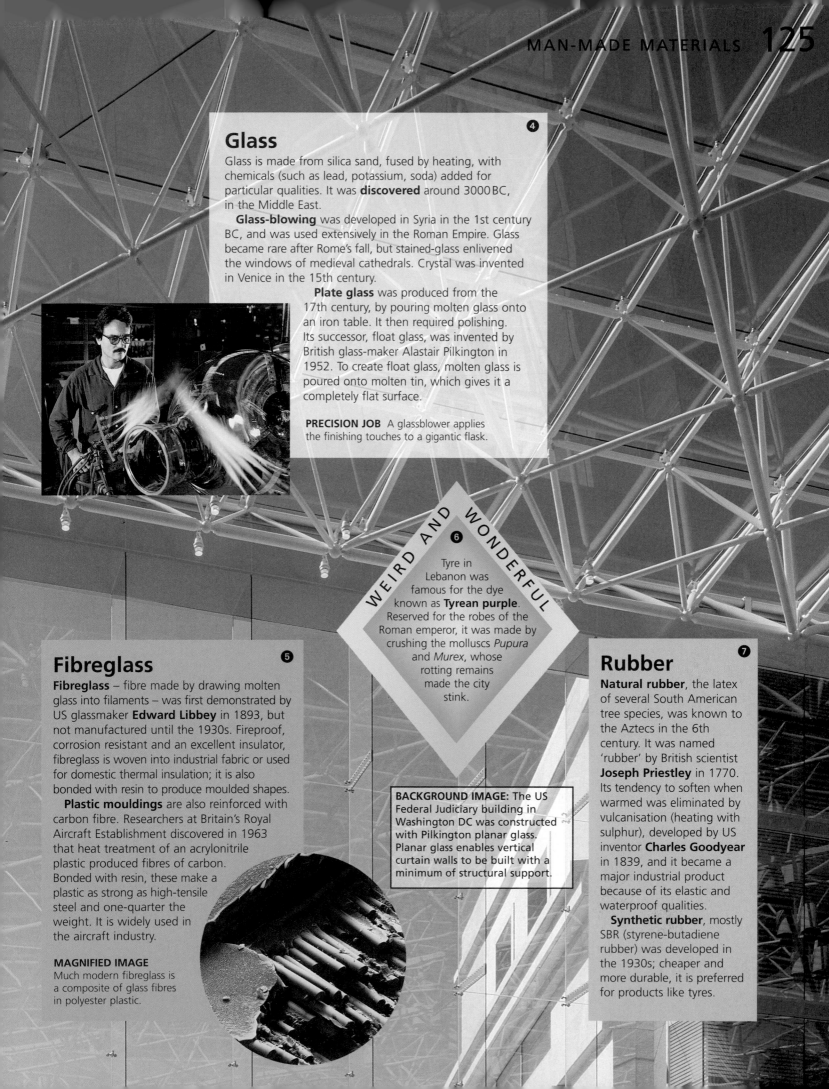

126

QUESTION NUMBER

The numbers or star following the answers refer to information boxes on the right.

ANSWERS

131 **D: Stainless steel ❸ ❺**

404 **Carbon ❶ ❺**

408 **Metals ❺**

466 **Pewter ❺**

469 **Aluminium, nickel and cobalt** – used in magnets

525 **The Bronze Age ❸**

594 **Bronze ❺**

596 **Steel** (the Pittsburgh Steelers) ❶ ❸

654 **Brass** – made up of copper and zinc

703 **A: Aluminium ❹**

708 **A: An iron-nickel alloy ❸**

743 **Mercury** – often combined with tin or silver

★ 816 **Copper ★**

818 **The philosopher's stone** – also linked to immortality

850 **Anneal** – also used to toughen glass

901 **Eat with it** (electro-plated nickel silver) ❷

975 **Coins** (and medals) ❺

Man-made metals

Smelting ❶

Smelting is the process of obtaining a metal from its ore by **heating** the ore **beyond the metal's melting point**. Smelting usually requires the presence of an oxidising agent – air, for example – or a reducing agent, such as carbon, to cause the necessary chemical changes. The waste products from smelting form a mix known as **slag**.

    Iron smelting requires very high temperatures. It was first practised around 1500 BC by the Hittites (who occupied Anatolia in modern Turkey), by heating iron ore with wood, which burned down to carbon-rich charcoal. The slag was separated from the metal by repeated hammering, resulting in **wrought iron**. **Cast iron** was produced in Europe from the 14th century AD as blast (forced air) furnaces were developed. Modern furnaces produce up to 8000 tons of iron every 24 hours, most of which is immediately converted into steel.

Iron ore, coke and limestone are dumped into a blast furnace.

Molten iron

Oxygen

Slag notch

Converter

Coke and air fuel combustion, producing molten iron that is full of impurities (pig iron).

To produce steel, molten iron is mixed with oxygen to burn off impurities.

The converter is tipped and molten steel is removed.

Electroplating and electrolysis ❷

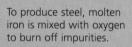

Electrolysis is the process of causing chemical change by passing an electric current through a solution; typically a metallic salt. The current causes positive ions (see page 118) in the solution to migrate to the negative electrode (**cathode**), and negative ions to the positive electrode (**anode**). The ions react with the electrodes, producing gas or deposited solids.

    Electroplating uses the process to coat one metal with another, for example on cutlery and jewellery. The object to be plated is dipped in a solution of metallic salt, such as silver nitrate, and a current is passed through the solution. The object acts as the cathode, and the plating metal is evenly deposited on its surface.

★ 816

Sterling work

Pure silver is too soft to be used for jewellery so other metals are added to it to make it workable. **Sterling silver** is only 92.5 per cent pure; the other 7.5 per cent is copper.

    Silver-plated jewellery is made from inexpensive base metals, such as brass, covered with a very thin layer of pure silver.

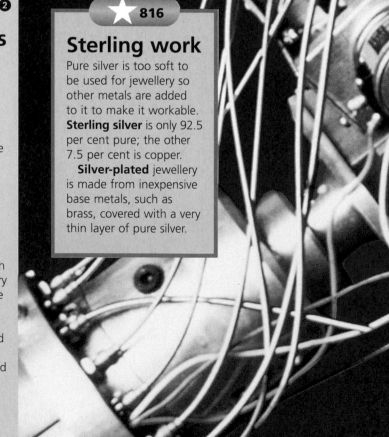

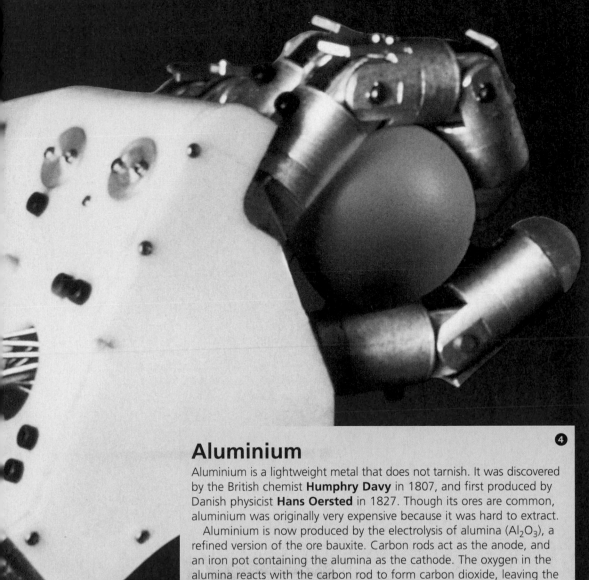

TIMESCALE ❸

▶ **3500 BC** Bronze, an alloy of copper and tin that is harder than either, is first made in the Middle East.

▶ **c.1500 BC** Iron smelted by the Hittites of Asia Minor.

▶ **c.1400 BC** Brass (alloy of copper and zinc) is made.

▶ **c.600 BC** Cast iron is first produced, in China.

▶ **c.300 BC** Steel is developed by the Romans.

▶ **c.AD 700** The Catalan forge, an early blast furnace, is developed in Spain.

▶ **1400** Improved furnaces make cast iron in Europe.

▶ **1709** England's Abraham Darby introduces coke as a high-carbon fuel for blast furnaces.

▶ **1784** Englishman Henry Cort designs the reverberatory furnace for high-quality wrought iron.

▶ **1839** In the USA, Isaac Babbitt invents bearing metal, an alloy of tin, copper and antimony.

▶ **1856** England's Henry Bessemer introduces a process for making cheap steel from pig iron produced by blast furnaces.

▶ **1890** Switzerland's Charles Guillaume develops Invar, an iron-nickel alloy that expands little when heated.

▶ **c.1900** Frederick Taylor (USA) develops tool steel.

▶ **1913** Englishman Harry Brearley invents stainless steel; in 1914 Germany's Krupps company produces an improved version.

▶ **1939-45** Second World War prompts development of the high-strength, heat-resistant Nimonic alloys of steel (with molybdenum, cobalt or tungsten) suitable for jet engines.

Aluminium ❹

Aluminium is a lightweight metal that does not tarnish. It was discovered by the British chemist **Humphry Davy** in 1807, and first produced by Danish physicist **Hans Oersted** in 1827. Though its ores are common, aluminium was originally very expensive because it was hard to extract.

Aluminium is now produced by the electrolysis of alumina (Al_2O_3), a refined version of the ore bauxite. Carbon rods act as the anode, and an iron pot containing the alumina as the cathode. The oxygen in the alumina reacts with the carbon rod to form carbon dioxide, leaving the metal. Alumina (or 'corundum') is also used as the abrasive emery.

Alloys and their uses ❺

Coinage bronze An alloy of copper, tin and zinc; widely used for 'copper' coins.

Coinage silver An alloy of silver with copper and lead; long used for 'silver' coins.

Dental gold An alloy of chromium, molybdenum, cobalt and titanium; used as an alternative to pure gold.

Duralumin A strong, extra-lightweight alloy of copper with aluminium, magnesium and manganese; used in aircraft construction.

Nichrome The trade name for a range of alloys of nickel and chromium, with small amounts of magnesium, iron, carbon or silicon. Nichrome has a high melting point and is corrosion-resistant. It is used in heating elements.

Manganin An alloy of copper, manganese and nickel with a high resistance to electric current. Manganin is used primarily in electrical resistors for circuit boards.

Pewter Alloys of tin with lead, copper or antimony; once widely used for domestic utensils.

Solder A general name for alloys used to join other metals together by melting. Soft solders (tin and lead) are used for joining copper wires. Hard or brazing solders, such as silver solder (copper, silver and zinc), form a stronger joint.

Stainless steel An alloy of iron, chromium and nickel, stainless steel; first produced in England in 1913. Perhaps the most familiar of all alloys, it is used particularly for cutlery and kitchen fittings.

Tool steel A general name for any hard steel with between 0.9 and 1.4 per cent carbon; used for tools that cut or work metal.

Man-made polymers

TIMESCALE ❶

▶ **1862** British chemist Alexander Parkes makes the first plastic, Parkesine, from cellulose.

▶ **1869** US chemists John and Isaiah Hyatt develop Celluloid.

▶ **1890** Hilaire de Chardonnet (France) manufactures rayon from cellulose.

▶ **1907** The first all-synthetic plastic, Bakelite, is developed by Belgian-born chemist Leo Baekeland.

▶ **1922** German chemist Hermann Staudinger discovers that the petroleum product styrene forms a polymer when heated.

▶ **1929** German chemical company IG Farben patents polystyrene.

▶ **1930** Perspex (called Plexiglas in the USA) is developed in Britain, Germany and Canada.

▶ **1931** Wallace Carothers of DuPont invents neoprene, a strong synthetic rubber.

▶ **1933** British chemist Reginald Gibson invents polyethylene (polythene).

▶ **1934** The first fully coated cellophane adhesive tape (manufactured by 3M) is marketed in the UK.

▶ **1937** The first successful all-synthetic fibre (nylon) is developed, by Wallace Carothers of DuPont.

▶ **1938** Roy Plunket of DuPont discovers Teflon.

▶ **1941** Polyester is developed by British chemists James Dickson and Rex Whinfield.

▶ **1943** German scientists create plasticised PVC as a substitute for rubber in electrical insulation.

▶ **1950** Acrylic plastic is first produced (marketed as Orlon in the USA from 1951).

▶ **1959** Lycra is developed and marketed by DuPont.

Cellulose to cellophane ❷

Cellulose is a natural carbohydrate that forms the structure of most plants. Cellophane is a film of cellulose, first made by the Swiss chemist **Jacques Brandenburger** in 1908. Wood pulp is dissolved in sodium hydroxide and carbon disulphide. The resulting solution is forced through a slit into an acid bath, where it settles into a transparent film.

FASHIONABLE FIBRE By the time of this advert for Morley stockings in the 1950s, nylon had become the height of glamour.

Plastics ❸

The word 'plastic' is derived from the Greek *plastikos*, meaning 'mouldable'. The chemical structure of plastics consists of long molecular chains that are formed in a process known as polymerisation.

The most important types of plastic include **thermoplastics**, which soften when warmed (for example polystyrene, polythene, and polyvinyl chloride or PVC); **thermosets**, which remain hard once set (such as Bakelite, epoxy resins and polyurethane); **polyamides**, used for producing sheets, films or fibres (such as nylon); and **silicones**, which are chemically inert and are used in insulation and plastic surgery.

NO WEAKEST LINK At the molecular level, plastics appear as long chains of identical units.

Recent developments include **shape-memory polymers** (such as polyisoprene), which can be deformed but resume their moulded shape when heated; and **biodegradable** plastics, which are largely made of carbon dioxide and water, or (like polyhydroxybutyrate or PHB) from sugars.

Acrylic paint ❹

All paints are composed of two parts, the **pigment** and the **binder**. Historically, paint was made from pigments bound with oil, natural glues or beeswax. In the 1940s, chemists added pigment to a water-soluble acrylic polymer binder, and acrylic, or emulsion, paint was born. Today, the majority of paint produced for industrial and domestic use is acrylic-based.

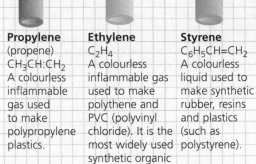

SUB-AQUA SNAPPER
This underwater camera has a tough polycarbonate casing to protect it from pressure at depth.

Oil derivatives ⑥

Converting crude oil into petroleum results in many by-products. Originally treated as waste or simple fuels, these petrochemicals are now used to make a huge range of chemical products. They include:

Propylene
(propene)
$CH_3CH:CH_2$
A colourless inflammable gas used to make polypropylene plastics.

Ethylene
C_2H_4
A colourless inflammable gas used to make polythene and PVC (polyvinyl chloride). It is the most widely used synthetic organic compound.

Styrene
$C_6H_5CH=CH_2$
A colourless liquid used to make synthetic rubber, resins and plastics (such as polystyrene).

Tie-breaker ⑤

Q: What does the 'poly' in polymer mean?
A: Many (from the Greek *polloi*). Polymers are made up of long chains of identical units. Polystyrene, for example, consists of lots of styrene ($C_6H_5CH=CH_2$) molecules joined together. The suffix '-mer' simply means part or share.

BACKGROUND IMAGE
Nylon fibres are waterproof but do not normally 'breathe'. The nylon fibres shown here, magnified 160 times, are coated with a porous synthetic rubber to make 'breathable' fabric for more comfortable protection from the rain.

NO WRINKLES The crease-resistant fabric Crimplene was popular in the 1960s and 1970s.

Man-made fibres ⑦

Lycra
A polyurethane fabric that is much stronger and more elastic than rubber-based alternatives. It is used chiefly in underwear and sportswear.

Kevlar
A plastic five times stronger than steel, invented in 1971 by Stephanie Kwolek of DuPont. It is used in tyres, cables, armour and safety helmets.

Nylon
A polyamide notable for its strength, elasticity and resistance to abrasion and chemicals. Nylon is widely used for fabrics and thread, but also ropes, and solid mouldings such as gears.

Orlon
An acrylic fibre that combines bulk with light weight and resistance to fading. It is used for jumpers and other clothing.

Rayon
The name for a range of synthetic fibres made from cellulose. It is produced in greater quantities than any other fibre. Viscose rayon is widely used for clothing, furniture and carpets.

Terylene
(Dacron in the USA.) Polyester with a high tensile strength and resistance to stretching. The yarn is used in curtains, clothing and thread – the staple fibre mixed with wool in clothing.

★ 536

Glues

An adhesive is any substance that sticks two surfaces together. Natural adhesives (glues) include bitumen (first used 40 000 years ago), beeswax and gelatins from animal skin, bone and sinew. Modern adhesives, including epoxy resins and superglues, are far stronger.

WEIRD AND WONDERFUL ⑧
Plastics have one major disadvantage: they do not decompose. In 2002, the Irish government introduced a 'tax' on plastic bags to limit the build-up of waste. Today Ireland's shoppers pay an extra 10p for every plastic bag they use.

Cycles of life

Core facts ❶

◆ **All living things** share certain characteristics which can be used to differentiate them from the non-living. **Viruses**, however, are difficult to classify, and lie on the borderline between living and non-living things.

◆ Two universal requirements among living things are for the elements **carbon** and **nitrogen**, both of which are used and recycled constantly through the environment.

◆ **Plants** and **animals** both form part of the web of interacting life forms on Earth. But while plants could survive without animals, animals could not survive without plants – plants provide animals with food and oxygen.

What is life? ❷

Six processes and characteristics occur in all living things:

◆ **Adaptation** In the short term, living things adapt to environmental changes by changing their behaviour or growth patterns; in the longer term they adapt by evolution.

◆ **Growth** Organisms show ordered, controlled structural growth – plants grow continuously, most animals until maturity.

◆ **Metabolism** Two types of chemical process take place in all cells: the breakdown of substances to release energy; and the creation of complex substances, such as proteins in animals.

◆ **Movement** Most animals can move about freely; plants move by growth.

◆ **Reproduction** Sexual (mating) or asexual (budding) reproduction ensures the continuation of the species.

◆ **Responsiveness** (or irritability) All living things respond to external stimuli.

On the edge ❸

Viruses show many, but not all, characteristics of life. Unlike plants and animals, they have **no cell walls** and **cannot move** themselves. Viruses contain genetic material (see page 134), and **can reproduce**, but only by taking over part of the machinery of living cells in organisms such as plants or animals. Some scientists consider viruses as the smallest living things, but to the majority they are highly complex, self-replicating, non-living chemicals.

DEADLY ENEMY The HIV virus that causes AIDS.

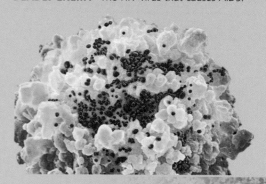

Cycles of life

Carbon and nitrogen are essential to all life forms. They are recycled between the land, oceans, atmosphere and living things.

Carbon dioxide is released into the atmosphere by living things through respiration. It is taken up by plants in photosynthesis, and oxygen is released. The simplest link in the carbon cycle is when an animal eats a plant.

Nitrogen, in the form of inorganic compounds in soil, is absorbed by plants. Some plants are eaten by animals, which may in turn be eaten by carnivores. Nitrogen returns to the soil in manure.

NITROGEN CYCLE Before nitrogen can be used by plants and animals it has to be 'fixed' in the form of chemical compounds known as nitrates. ❹

Nitrogen composes 78 per cent of the Earth's atmosphere, but it cannot be used directly by animals.

Lightning 'fixes' some atmospheric nitrogen as nitrogen oxide, which falls to Earth dissolved in rain.

Nitrogen-fixing bacteria in the sea, the soil and the root nodules of some plants convert nitrogen into nitrates.

Other bacteria convert some nitrates into nitrogen gas, released back into the atmosphere.

How plants make food ❺

Photosynthesis takes place in all green plants. It is the most important process for maintaining life on Earth, creating food for animals as well as plants.

Photosynthesis depends on the green substance **chlorophyll**, which is found in tiny structures called chloroplasts in leaves. Chlorophyll absorbs light energy from the Sun, and **chloroplasts** use this energy to synthesise the sugar **glucose**, a simple carbohydrate, from atmospheric carbon dioxide and water from the soil. The 'waste product' of photosynthesis is **oxygen**, which is released into the air. Animals and plants both use this oxygen to break down food materials – a process called **respiration** – releasing energy to power life processes.

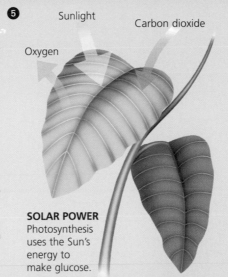

Sunlight
Carbon dioxide
Oxygen

SOLAR POWER Photosynthesis uses the Sun's energy to make glucose.

FOOD FACTORY Every cell in a leaf has at least one chloroplast (below), each of which is filled with stacks of thylakoids containing chlorophyll.

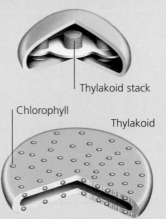

Thylakoid stack

Chlorophyll

Thylakoid

INSIDE A LEAF The background image shows the leaf of a pea plant seen though an electron microscope. The structures on the right are chloroplasts.

Respiration ❽

Respiration is not simply breathing, the technical term for which is external respiration. What scientists call respiration takes place inside living cells in structures called **mitochondria**. It is a process that releases energy from food. Glucose (made by breaking down more complex foods) is combined with oxygen to form carbon dioxide and water, which are excreted. In the process, energy-charged molecules called ATP are formed. These molecules energise all the processes of **metabolism**, enabling animals and plants to live and grow.

WEIRD AND WONDERFUL ❼

Two **beetles** are among many small creatures vital for maintaining cycles of life. Sexton beetles bury dead animals as food for their larvae – and thus ensure that the carcasses rot. Dung beetles do the same with animal droppings.

Life in miniature ❻

A Dutch draper and amateur scientist, **Anton van Leeuwenhoek** (1632-1723) discovered remarkable things using simple single-lens microscopes he made, which could magnify 200 times.

He was the first to see, draw and describe bacteria, protozoa and other micro-organisms, which he called 'animalcules'. His findings helped to disprove the idea of the spontaneous generation of lower life-forms.

NEW WORLD A page from Leeuwenhoek's letters to the Royal Society in London with drawings of bacteria.

⭐ 58

One-world theory

The **Gaia hypothesis** was originated in 1968 by the British scientist **James Lovelock**. It suggests that the planet's biosphere (all living things) and its physical environment act together like a single huge living organism. Rather than seeing living things as engaged in a continuous battle for survival in their environment, it suggests that the Earth is self-regulating, capable of adapting to keep itself fit for life to inhabit.

The structure of living things

Core facts ❶

◆ All truly living organisms consist of units called **cells** – the smallest structures that can carry out all the functions of life.
◆ Some organisms consist of single cells, but most of the plants and animals that we are familiar with are multicellular – in other words they have many cells.

◆ Most cells are too small to see without a microscope; the human body contains about 10 million million of them.
◆ An intricate web of chemical reactions takes place inside all cells. The most important types of chemicals in these reactions are **proteins**, **carbohydrates**, **lipids** and **nucleic acids**.

Cell structure ❷

All true cells have a membrane separating them from the outside world and enclosing jelly-like **cytoplasm** inside. They have tiny protein-building bodies called **ribosomes** and the genetic material **DNA** (see page134). In all cells except bacteria, most of the DNA is in the **nucleus**, and there are various other specialised structures, or **organelles**. These include:
◆ **Endoplasmic reticulum**, a network of channels where fats are synthesised and ribosomes found.
◆ **Lysosomes**, which break down toxins and other unwanted substances.
◆ **Mitochondria**, where food materials are 'burned' to release energy for life processes.

Chemical basis for life ❸

All living organisms are at least 60 per cent water. Other important categories of chemicals include:
◆ **Proteins** The main structural material of animals' bodies. Proteins also include **enzymes**, which control all biochemical activity. Proteins are built from chains of units called **amino acids**, which contain nitrogen, carbon, oxygen and hydrogen atoms, and in some cases sulphur.
◆ **Carbohydrates** Made of carbon, hydrogen and oxygen atoms, these include cellulose – the structural material of plants – starch and sugars. Starch and sugars supply energy to organisms.
◆ **Lipids** These also contain carbon, hydrogen and oxygen. Lipids include oils, fats, phospholipids (which form cell membranes) and steroids (including some hormones).
◆ **Nucleic acids** Known by their initials DNA and RNA, these control heredity and protein synthesis (see page 134).

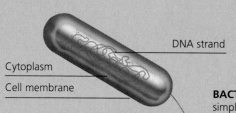

Cytoplasm
Cell membrane
DNA strand

BACTERIUM Bacterial cells (left) have a simple strand of DNA and no other internal structures apart from ribosomes.

PLANT CELL A green plant's cell (right) is unique in having chloroplasts (which contain chlorophyll and are the site of photosynthesis), a tough cell wall and a sap-filled central vacuole. The vacuole keeps the cell rigid.

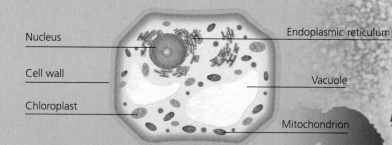

Nucleus
Cell wall
Chloroplast
Endoplasmic reticulum
Vacuole
Mitochondrion

ANIMAL CELL Having no cell wall or vacuole, a typical animal cell (right) is irregular in shape.

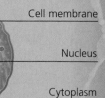

Mitochondrion
Cell membrane
Nucleus
Cytoplasm

Cells, tissues and organs ④

All cells have the same basic components, but they vary in shape, form and function. A collection of similar cells with a specific function is called **tissue**; groups of tissues form **organs**. There are four main types of **animal tissue**:
◆ **Connective tissue** connects and supports parts of the body, and includes non-cellular material – bone, blood, fat.
◆ **Epithelium** contains tightly packed cells, often in sheets, and forms membranes, skin and glands.
◆ **Muscle** consists of elongated cells that can contract.
◆ **Nerve tissue** makes up the brain and nervous system.
Plant tissues are divided into **meristematic** (grows continuously, forming root and shoot tips) and **permanent** (mature cells). Permanent plant tissues include the epidermis (outer covering), parenchyma (inner parts of leaves and other organs), phloem and xylem (internal transport tissues).

BUILT FOR THE JOB White blood cells (leucocytes) fight infection by engulfing invading organisms (right). **BACKGROUND** Nerve tissue is made up of cells with long projections (axons) along which information is transmitted from cell to cell.

The life scientists ⑤

◆ **Aristotle** (384-322 BC) analysed animal parts in terms of their purpose.
◆ **Pliny the Elder** (c.AD 23-79) wrote a 37-volume *Natural History*, including details of many plants and animals.
◆ **Galen** (c.129-210), a doctor, founded comparative anatomy.
◆ **Leonardo da Vinci** (1452-1519) made detailed anatomical drawings.
◆ **Andreas Vesalius** (1514-1564) wrote the first anatomy textbook.
◆ **William Harvey** (1578-1657) discovered the circulation of blood.
◆ **Marcello Malpighi** (1628-94), **Anton van Leeuwenhoek** (see page 131) and **Robert Hooke** (1635-1703) were among the first to use microscopes to make biological observations.
◆ **Carolus Linnaeus** (1707-78) founded the modern system of classifying and naming organisms.
◆ **Baron Cuvier** (1769-1832) founded palaeontology (the study of fossils).
◆ **Charles Darwin** (1809-82) and **Alfred Russel Wallace** (1823-1913) independently originated the theory of evolution by natural selection.
◆ **Matthias Scheilden** (1804-81) and Theodor Swann (1810-88) showed cells are the basic units of living things.
◆ **Gregor Mendel** (1822-84) discovered the laws of heredity.
◆ **Louis Pasteur** (1822-95) and **Robert Koch** (1843-1910) founded bacteriology.
◆ **Thomas Hunt Morgan** (1866-1945) showed that chromosomes carry genes.

The simplest organisms ⑥

Many living things exist as single independent cells. Most are microscopically small, and they include the simplest organisms of all: **bacteria** and **blue-green algae**, which are classified in their own kingdom, Prokaryotae.

Most other single-celled organisms – including both animal-like types such as amoebas and other protozoans, and plant-like diatoms and other algae (seaweeds) – are usually grouped today in the kingdom Protista, or **protists**, separate from true plants and animals. A third group are yeasts, which are **single-celled fungi**.

★ 323

Dwarves and giants

Most animal cells average 0.01-0.02 mm across; plant cells are a little larger. Big organisms generally have more rather than bigger cells. But cells can vary greatly in shape and size. The **largest cells** of all are egg cells – about 0.1 mm in humans, and the size of a tennis ball in ostriches – only the yolk is the true cell. Brain cells may be as small as 0.005 mm, but some nerve cells are 1.2 m (4 ft) long, though far narrower than a hair.

134

Genetics and DNA

The numbers or star following the answers refer to information boxes on the right.

Core facts ❶

◆ Genetics – the study of **biological heredity** – has revolutionised both the life sciences and our potential to treat disease.
◆ Individual **inherited characteristics**, or traits, that are passed down the generations are governed by genes.

◆ **Genes** are found within the nucleus of each cell in structures called **chromosomes**, and consist of strands of a chemical called DNA.
◆ **DNA** is self-replicating and consists of units known as **bases**. The sequence of bases spells out a chemical code called the **genetic code**.

What is DNA? ❷

DNA stands for **deoxyribonucleic acid**. It is a complex chemical found in chromosomes in the **nucleus** (left) of all cells, and carries the instructions for making the proteins that a cell needs to function.
Proteins control an organism's characteristics. By governing protein production, DNA transmits inherited traits from generation to generation.

DNA consists of chains of paired units called bases. There are four bases: **adenine** (A), **cytosine** (C), **guanine** (G) and **thymine** (T). Adenine always pairs with thymine, and cytosine with guanine. The genetic code consists of series of three-letter 'words' – sequences of these bases. When cells divide, DNA replicates itself exactly, copying the coded message.

Adenine

Cytosine

Thymine

Guanine

DOUBLE HELIX
A DNA molecule has twin 'corkscrew' strands of sugars cross-linked by bases, which fit together like a lock and key. Base A always pairs with T, and C with G, so one strand forms a mould or template for making the other.

A tale with a twist ❸

A flash of insight in 1953 by Cambridge scientists **James Watson** (1928-) and **Francis Crick** (1918-) caused a revolution in genetics. Using X-ray diffraction images of DNA taken at King's College, London, by **Rosalind Franklin** (1920-58) working with **Maurice Wilkins** (1916-), they realised that DNA molecules must consist of two intertwined **helixes** (corkscrews).
It was known that DNA contains the same number of adenine units as thymine, and the same number of guanine as cytosine. Crick and Watson realised that the pairs of bases must be linked like rungs in a spiral staircase, and that this explained DNA's ability to replicate itself. Molecular genetics was born.

The Human Genome ❹

Human DNA contains about 3 billion pairs of bases (see above) and an estimated 100 000 or more genes. This complete genetic make-up of a person is called the genome, and the aim of the Human Genome Project is to work out the complete sequence of bases and genes for human beings.
Work started in Britain and the USA in 1990, and a 'rough draft' was completed in 2000. When the complete version is finished in about 2003, it should be possible to design new treatments for many hereditary and other diseases.

BACKGROUND IMAGE A computer display shows the sequence of bases in a fragment of human DNA. Green peaks are adenine, red thymine, yellow guanine and blue cytosine.

Dominant or recessive? ❺

Gregor Mendel (1822-84), an Austrian monk, discovered how creatures inherit traits (physical characteristics) through experiments on breeding pea plants. He found that traits are handed down from parents to offspring through hereditary 'factors' now called genes – one gene for a trait from each parent.

In the case of related traits, such as eye colour in humans, some are dominant (brown eyes) and others recessive (blue eyes). A dominant gene always shows up – a person with the brown gene always has brown eyes. A recessive gene is apparent only if it is inherited from both parents. If both parents carry the dominant and recessive versions (alleles) of a gene, the genes are mixed at random and some offspring will show the recessive trait, some not.

ALL IN THE EYES In this example, white eyes (W) are dominant and red eyes (r) recessive. One parent has white eyes (WW) and the other red (rr). Their offspring (the F1 generation) both inherit a dominant white gene and a recessive red gene (Wr), so both have white eyes.

If one of the offspring (Wr) mates with a fly with red eyes (rr), red eyes may appear again in the next generation (F2). Here, one out of four shows the recessive trait.

WW   **rr**

Parental generation: White eyes and red eyes.

Wr   **Wr**

F1 generation: Both offspring have white eyes.

Wr   **rr**   **WW**   **rW**

F2 generation: In the second generation from the original parents, three flies have white eyes and one has red eyes.

Genetic terminology ❼

◆ **Allele** One of the alternative forms (dominant or recessive) of a gene.
◆ **Chromosomes** Tiny structures in a cell's nucleus that contain DNA, seen when the nucleus is dividing to form new cells.
◆ **Gene** The inherited material that governs a physical or behavioural trait (characteristic) of an organism. Once studied purely theoretically, genes are now known to consist of a segment of DNA.
◆ **Genome** An organism's complete set of genes.
◆ **Heterozygote** An individual that inherited dissimilar alleles from its parents (as with the F1 generation flies in the diagram on the left).
◆ **Homozygote** An individual that inherited the same alleles from both parents (as with the black-eyed flies and white-eyed parent).
◆ **Phenotype** The physical effects of the expression of a particular gene.
◆ **RNA** Ribonucleic acid, a similar chemical to DNA. It forms the genetic material of some viruses. In other organisms it plays a vital role in protein synthesis by 'translating' the DNA code and assembling amino-acid units in the correct sequence, and in controlling the operation of genes.

Genetic engineering and cloning ❻

An understanding of the chemical nature of genes has enabled scientists to alter the genes an organism passes on to its offspring. This process of alteration is known as **genetic engineering**.

Genetic engineering can be used to remove genes that cause hereditary diseases, or to insert genes from a different organism to give an animal or plant resistance to disease or pesticides. In some cases, complete synthetic chromosomes have been inserted – such as human DNA into cows for producing antibodies.

Genetic engineering may be combined with **cloning** – the production of identical individuals by inserting extra nuclei into cells – but cloned organisms seem to age faster than normal individuals. The first cloned mammal was Dolly the sheep, born in 1996.

LUMINOUS NOSE Both of these piglets are clones but the one on the left has also been genetically engineered. Jellyfish genes added to its DNA result in its skin being yellow.

 ★ **409**

Deciding gender

Every species contains a specific number of chromosomes in each cell. The chromosomes are mostly arranged in pairs (one from each parent).

Most humans have a total of 46 chromosomes per cell. Women have 23 complete pairs, including two so-called **X chromosomes**. Men have only one X chromosome (from their mother), and a smaller **Y chromosome** (from their father). It is the Y chromosome that produces a male rather than female embryo, and it is passed on virtually unchanged from father to son.

A comparison of the DNA in human Y chromosomes can show how closely related men are. For instance, one study showed that most British men with the same unusual surname are related.

Agricultural revolution

Core facts ❶

◆ **Food** is one of the fundamental needs for human survival, so its **production** has been a vital activity ever since people ceased to be hunter-gatherers and settled in communities.
◆ There has been constant pressure to raise agricultural productivity. In 1850, the average American farmer produced enough to feed five people; today he grows food for 80.
◆ The main factors in increasing food output have been irrigation, mechanisation, the use of chemical fertilisers and pesticides, and the creation of new **crop** and **stock** varieties.

Modifying nature ❷

Since agriculture began, farmers have been **selectively breeding** stock and crops by choosing the best for the next generation.

Gregor Mendel's studies in heredity (see page 135) in the 19th century made sophisticated breeding programmes possible, leading to high-yielding, sturdy crops. Hybrids such as the broccoflower (a broccoli-cauliflower cross) were created in a similar way.

In 1994 the first commercial **genetically modified** (GM) crop, the 'Flavr Savr' tomato, was made by altering the plant's genes. However, GM crops still face problems and consumer resistance.

TRANSGENIC FRUIT Melon plants grown from genetically modified seed produce fruit with a much longer shelf life.

PRECISION FARMING The cockpit of a spray rig is equipped with computerised controls.

Mechanised farming ❹

Computer-controlled **combine harvesters** are the latest step in a process of high-volume mechanisation that began with Jethro Tull's horse-drawn **seed drill** of 1701.

In the 19th century, steam replaced horse-power in some farm machinery, but it was the internal combustion engine that revolutionised farming. The first **tractor** appeared in 1903, but the 1916 Fordson was the first mass-produced model. By the 1920s, a combine harvester could cut and thresh 0.8 ha (2 acres) of cereal crops an hour.

Today's trend is towards 'smart farming'. Specialised **land-survey satellites** monitor crop growth and signal precise fertiliser, pesticide and water needs and the best time to harvest.

Tie-breaker ❸

Q: Which major sheep-rearing country has more sheep per capita than any other?
A: New Zealand. With a sheep population of over 47 million, and a human population of around 3.8 million, it has more than $12\frac{1}{2}$ sheep per person. Only Australia comes near this record, with over six sheep per capita.

Fertilisers and pesticides ❺

The agricultural revolution of the last half-century was as much **chemical** as mechanical and genetic. **Artificial fertilisers**, first applied in the early 19th century, were widely used once German chemist Fritz Haber developed a process to make ammonia from air in 1913.

After the Second World War, short-stemmed, sturdy crops were bred that could be heavily fertilised but not fall over under their own weight. **DDT**, the first **insecticide**, was discovered in 1939 and 2,4-D, the first **selective weedkiller**, in 1945.

One aim of genetic research is to develop **GM crops** that resist weedkillers so that they can survive spraying that kills competing weeds.

Hydroponics ❻

Crop plants have to get all their **nutrients**, plus water, from the soil – which also anchors them. Even 'organic' fertilisers such as manure are broken down in the soil into simple chemicals before the plants can absorb and use them to build roots, leaves, flowers, fruits and so on. **Hydroponics** supplies all these nutrients directly in the form of chemicals dissolved in water, which flows over the plants' roots.

The plants grow in a loose, **inert material**, such as vermiculite, sand or lightweight artificial pebbles, which gives support without hindering root growth. Lighting and the supply of nutrients and trace elements are controlled automatically.

BACKGROUND IMAGE: Rows of lettuces growing under hydroponic conditions in a large-scale commercial greenhouse.

TIMESCALE ❼

ANCIENT SCENE An Egyptian funerary model shows ploughing around 2000 BC.

▶ *c.*8000 BC Crops first farmed in the Middle East.
▶ *c.*5000 BC First rice paddies in China.
▶ *c.*3500 BC Maize widely grown in the Americas. Ox-drawn wooden ploughs used in Mesopotamia.
▶ *c.*3000 BC *Shaduf* (water-lifting device) used for irrigation in Egypt.
▶ *c.*1000 BC Chinese invent curved mouldboard to turn ploughed soil.
▶ *c.*600 BC Chinese first use crop rotation.
▶ *c.*500 BC Chinese use iron ploughshare (cutter).
▶ *c.*300 BC Harrow first used for tilling soil.
▶ 3rd century BC Archimedes' screw used for irrigation.

▶ **4th century AD** Flail introduced for threshing in Roman world.
▶ *c.*800 Open (shared) field system in Europe.
▶ *c.*1100 Early ripening rice introduced in China.
▶ **16th century** Tomatoes, maize and potatoes introduced to Europe from Americas.
▶ **1701** Jethro Tull invents horse-drawn seed drill.
▶ **1730s** Norfolk 'four-course' crop rotation introduced in Britain.
▶ *c.*1760 Scientific stock breeding begins in Britain.
▶ **1786** Horse-driven threshing machine.
▶ **1808** First iron plough.
▶ **1830s** Reaper, horse-drawn reaper-thresher and steel plough invented.
▶ **1862** Milking machine.
▶ **1884** Combine harvester.
▶ **1916** Ford starts to mass-produce tractors.
▶ **1940s** Intensive rearing of poultry, pigs and dairy cattle.
▶ **1960s** 'Green revolution' brings high-yielding crops.
▶ **1990s** First genetically modified crops.

★ 219

Farmed fish

Although **fish farming** – raising fish in enclosures supplied with pelleted food – is increasing greatly, it is an ancient practice. People in eastern Asia have raised species such as milkfish in village ponds for centuries.

Irrigation ❽

The most basic need of crops is **water**, and nearly half of all today's food production, in terms of value, comes from **irrigated land**. The Ancient Chinese, and later the Egyptians, irrigated land thousands of years ago. Today's big irrigation schemes involve building vast dams to store flood waters, and may have bad environmental effects. It is as important to deliver water to crops with as little wastage from evaporation and runoff as possible as it is to collect and distribute it to where it is needed.

Food technology

Core facts ❶

◆ Most foods deteriorate sooner or later, due to the action of micro-organisms. Food technology has grown from **preserving** food to **processing** for flavour and texture and the creation of novel and '**convenience**' foods.
◆ Preservation involves stopping the growth of

micro-organisms with chemicals, dehydration, cold or sterile sealing in a container.
◆ Food **additives** include colourings, flavourings and texture enhancers – some synthetic, others extracted from foodstuffs.
◆ **Artificial** foods date back over a century.

Traditional methods ❷

INSTANT MASH Freeze-dried potato appeared in 1968.

All methods of food preservation depend on preventing the growth of bacteria, fungi and other micro-organisms that rot foodstuffs.
◆ **Drying** Micro-organisms cannot grow if the moisture content in food is below a certain level. Traditional methods are sun or air-drying, or heating over a fire or in a slow oven.
Freeze-drying – vacuum-drying pre-frozen food so that the water in it turns directly to vapour – was introduced in 1938 and first used for instant coffee.
◆ **Curing** In curing chemicals are used to preserve food. It includes **salting**, **pickling** with vinegar or alcohol, the use of sodium nitrate and nitrite, and **smoking**. These substances (including those in wood smoke) stop or slow the growth of micro-organisms; **salt** also dries food. **Fermentation** partly decays food, but the process also releases preserving chemicals.

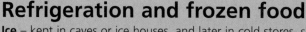

★ 149

Birdseye

Clarence Birdseye started thinking about frozen foods while working as a fur trader in Labrador. By 1925 he had perfected a process for **quick-freezing** fish between two refrigerated metal plates. The fish retained more flavour this way than with other methods.
Birdseye extended the process to fruit and vegetables. His name became famous after his company was taken over by what later became General Foods.

Refrigeration and frozen food ❸

Ice – kept in caves or ice houses, and later in cold stores, ships, railway wagons and home ice boxes – has been used to keep food fresh since at least 1000 BC. Mechanical **refrigeration systems** rely on the cooling produced by the evaporation or expansion of a fluid.
◆ **1834** Jacob Perkins invents the first refrigerator using the evaporation of compressed ether.
◆ **c.1845** John Gorrie invents a refrigeration system using the rapid expansion of compressed air.
◆ **c.1860** The Carré brothers in France invent a system using the evaporation of liquid ammonia.
◆ **1873** Karl von Linde builds the first practicable small-scale compression refrigerator.
◆ **1905** An absorption-type refrigerator, without a pump, is invented in Sweden.
◆ **1920s** Home refrigerators become common, using ammonia, methyl chloride or sulphur dioxide and, from around 1930, non-toxic chlorofluorocarbons (CFCs). CFCs' ozone-destroying property leads to new refrigerants being introduced from the 1980s.
◆ **c.1925** Clarence Birdseye develops quick freezing.

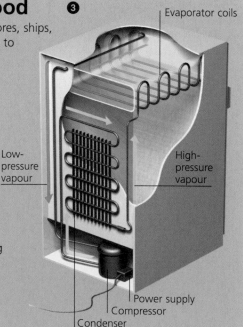

Evaporator coils

Low-pressure vapour

High-pressure vapour

Power supply
Compressor
Condenser

BIG MAC McDonald's is the largest fast-food retailer in the world, with more than 30 000 restaurants in 121 countries.

Fast food

When the **Earl of Sandwich** (1718-92) asked for a slice of meat between two pieces of bread, so he could eat without leaving the gaming table, he invented fast food as the sandwich. Today's fast food is a US invention. The first **hamburger chain**, White Castle, began selling meat patties in a bun in 1921, but the biggest fast-food companies were born in the 1950s. **Colonel Harlan Sanders** opened his fried chicken shop in Corbin, Kentucky, in 1952, and **Burger King** began in 1954 in Miami. Maurice and Richard McDonald opened an 'assembly-line' hamburger restaurant at San Bernardino, California, in 1940, but it was only in the 1950s that Ray Kroc started to expand **McDonald's** into a global empire. Frank and Dan Carney, using $600 borrowed from their mother, started **Pizza Hut** in Wichita, Kansas, in 1958.

LIGHT WEIGHT About 50 per cent of all food items consumed on the space shuttle are dehydrated. The water used to rehydrate the food comes from the fuel cells. Astronauts can choose between about 100 different food items and 50 beverages.

New and synthetic foods

◆ **Margarine** A French chemist, Hippolyte Mège-Mouriès, developed margarine in 1869. He called it 'oleomargarine' because the main ingredients were oleo (an oil from beef fat) and a fatty acid, margaric acid. Most modern margarines contain vegetable oils.

◆ **Artificial flavourings** The first synthetic version of a natural flavour was ethyl vanillin (the flavouring in **vanilla**), in the 1870s. **Monosodium glutamate** (MSG) was discovered in Japan in 1908. It occurs naturally in seaweed, and is used to enhance the flavour of a range of foods.

◆ **Artificial sweeteners** The first was **Saccharin**, found accidentally in 1879. It is about 300 times sweeter than sugar. More recent synthetic sweeteners include aspartame, cyclamates, sucralose and acesulfame potassium. The safety of some sweeteners has been questioned, and some are banned in certain countries.

◆ **Meat substitutes** Among the vegetable products processed to resemble meat are **textured vegetable protein** (TVP), a relative of tofu made from soya beans; and mycoprotein (sold as **Quorn**), made from a fungus, *Fusarium graminearum*, grown in fermentation tanks.

◆ **Processed ground fish** Surimi – usually seen as artificial crabsticks – and kamaboko, seen in loaf or cake form, are both long-established Asian foods made by mincing and processing white fish.

SPORTS DRINK The modern trend towards healthy foods emphasises 'natural' ingredients over synthetic additives.

Canned food

Italian naturalist Lazzaro Spallanzani found in the 18th century that meat extracts sealed in glass flasks and heated for an hour would keep for weeks. Nicolas Appert, in France, developed a commercial **bottling** process around 1804 in which sealed glass jars of food were heated in boiling water. '**Tin**' cans (in fact tin-plated steel) began to be used about ten years later. A scientific basis for the process emerged with Louis Pasteur's discovery in the 1860s that heat kills the **micro-organisms** that cause food spoilage; and, more importantly, with the discovery in 1896 of the bacterium *Clostridium botulinum*, the cause of deadly botulism. This led to modern **sterile canning** processes.

BOVRIL BOTTLE A glass Bovril container c.1886.

Medicine 1

Core facts ❶

◆ Over the past 500 years, we have made massive strides in our understanding of how the **human body** works and of **infections**.
◆ The underlying principle behind much medicine is to diagnose what is wrong, and what causes it, then choose the best treatment.
◆ A wide range of **techniques**, **tests** and

machines aid **diagnosis** today, enabling doctors to look inside the body.
◆ Treatments range from **drugs** to **surgery** and may involve medical **machines**, **transplants** or artificial **implants**.
◆ **Preventive medicine** is the most efficient form of treatment.

Studying the human body ❷

Most ancient religions forbade human **dissection**, but in India in the 8th century BC the surgeon **Sushruta** managed to study anatomy by pulling apart – rather than cutting – decaying dead bodies.

From AD 1300 in Europe, dissection (often of executed criminals) became part of medical training. Later milestones were Vesalius's first **anatomy** text in 1543, Harvey's discovery of **blood circulation** in 1628, and the invention of the **microscope** *c*.1590. From 1895, X-rays enabled doctors to 'see' inside the body.

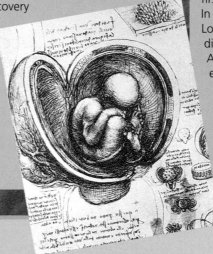

HUMAN FOETUS
A drawing by da Vinci.

Understanding infections ❸

For centuries, infections were blamed on miasmas, or noxious gases, from rotting matter and foul water. Then, in the 17th century, **micro-organisms** were first seen with a microscope. In the 1850s the French chemist Louis Pasteur linked these to disease with his **'germ' theory**. A German doctor, Robert Koch, established in the 1870s and 1880s that **bacteria** cause diseases such as anthrax and tuberculosis. **Viruses** were detected in the 1890s but not seen directly until the 1940s, with electron microscopes.

TIMESCALE ❹

▶ *c*.**8000 BC** Trepanning (skull-boring) practised.
▶ *c*.**3000 BC** First Egyptian and Chinese medical texts.
▶ **2000 BC** Early Indian medical text, *The Vedas*.
▶ **8th century BC** Indian medical text, *Sushruta-samhita*. Cataract operations in India.
▶ *c*.**400 BC** Hippocrates notes association of malaria with certain localities.
▶ *c*.**AD 170** Galen proves arteries and veins carry blood.
▶ **1543** Vesalius writes first illustrated book of anatomy.
▶ **1628** William Harvey discovers blood circulation.
▶ **1694** Bacteria first seen with microscope.
▶ **1726** Stephen Hales measures blood pressure.

▶ **1753** James Lind uses citrus fruits against scurvy.
▶ **1780s** Digitalis used to treat oedema (dropsy).
▶ **1796** Edward Jenner vaccinates against smallpox.
▶ **1805** Morphine extracted (from opium).
▶ **1815** First stethoscope.
▶ **1842** Ether first used as anaesthetic.
▶ **1857** Pasteur's 'germ' theory of disease.
▶ **1865** Joseph Lister pioneers antiseptic surgery.
▶ **1882** Robert Koch finds tuberculosis bacterium.
▶ **1895** First use of X-rays.
▶ **1899** Isolation of aspirin.
▶ **1900** Karl Landsteiner discovers blood groups.
▶ **1903** Electrocardiograph.

▶ **1910** Salvarsan – the first synthetic antibacterial drug.
▶ **1921** Insulin isolated and used to treat diabetes. First sticking plaster.
▶ **1928** Alexander Fleming discovers penicillin.
▶ **1929** EEG machine.
▶ **1938** First artificial hip.
▶ **1944** Open-heart surgery.
▶ **1953** First use of heart-lung machine in surgery.
▶ **1954** First polio vaccine.
▶ **1950s** Ultrasound imaging.
▶ **1967** First CAT scanner. First heart transplant.
▶ **1974** First MRI scanner.
▶ **1980** Smallpox eradicated.
▶ **1981** First use of artificial skin to treat burn patient.
▶ **1990s** Computer design of drugs becomes routine.

THE INNER MAN
A magnetic resonance
imaging (MRI) scan of
a man's body is set
against a background
of a colon X-ray.

ABO and Rh

The discovery of the blood groups
A, B, AB and O by Austrian-born
Karl Landsteiner in 1900 has saved
millions of lives by enabling safe blood
transfusions after accidents, injury or
surgery. Before the cross-matching of
blood groups, blood clots or even
death often followed transfusions.
The first voluntary blood-donor scheme
started in London in 1921. Landsteiner
and a colleague discovered the
separate rhesus (Rh) positive and
negative groups in 1940.

Diagnostic tools   ❺

Doctors base treatment of diseases and
disorders on accurate diagnosis. They may
use computer-assisted machines to detect
abnormal changes in body structure,
function or chemistry, as well as more
traditional methods.

Basic techniques include feeling for
lumps and tenderness and observing
colour changes and rashes.
◆ The **stethoscope**, for
listening to heart and lungs,
was invented by René
Laënnec in 1815.
◆ The **clinical
thermometer** was invented
by Sir Thomas Allbutt in 1867.
◆ The first **sphygmomanometer**, for
checking blood pressure, was used in 1860,
but in 1896 Riva Ricci introduced the
modern type, which works by measuring
the pressure of air in an inflatable cuff
around the arm.

Examining inside the body was for a
long time possible only by dangerous
exploratory surgery.
◆ **X-rays** were first used to make a clinical
image by Wilhelm Röntgen in 1895.
◆ **Ultrasound imaging** was introduced in
the 1950s; it detects echoes of high-
frequency sounds from internal organs.
◆ **Endoscopy** enables doctors to see (and
operate) inside the body without major
surgery, usually with a fibre-optic or rod-
lens device, or with a tiny video camera.
◆ **Thermography**, a temperature-mapping
system that can show local inflammation,
was introduced c.1960.
◆ In computerised axial tomography (**CAT**)
scanning, a computer combines cross-

INSIDE THE ABDOMEN In this coloured X-ray
of the abdomen, an endoscope winds through
the colon; bones of the spine are shown as green.

sectional X-rays to make a 3-D image.
◆ Magnetic resonance imaging (**MRI**) uses
the interaction of radio waves with
hydrogen atoms in the body of a person,
placed in a strong magnetic field, to show
the structure of organs.
◆ Positron-emission tomography (**PET**)
scanning reveals chemical activity in
organs, including brain changes during
various activities.

Measuring electrical activity is among
the most important tools for heart and
brain specialists.
◆ The electrocardiogram (**ECG**) was
invented in 1903 by Dutch physiologist
Willem Einthoven, and traces electrical
activity of the heart – picked up by
electrodes on the skin – onto paper.
◆ The electroencephalograph (**EEG**) does
the same with brain activity; it was
invented in 1929 by Hans Berger.

Examining body tissues is the job of
histologists, haematologists (blood experts),
pathologists and others.
◆ The **microscope** is used by histologists,
looking for tissue changes, and by
pathologists, looking for micro-organisms
and other signs of disease.
◆ **Blood and urine tests** include
microscopic and chemical examination, for
many diseases change the chemical
content of body fluids. Blood may be
separated into its components in a rapidly
spinning **centrifuge** before examination.

Medicine 2

Drugs ❶

Early drugs were derived from herbal recipes: some of these were no more than placebos. Treatment using chemical substances is known as chemotherapy.

◆ **Natural medicines** in current use include analgesics (painkillers) such as aspirin (originally derived from willow-tree bark) and morphine (from the opium exuded by poppy seed-heads); digitalis (from foxgloves), used to regulate the heartbeat; and insulin to treat diabetes (formerly extracted from the pancreas of cattle and pigs, now made by genetically engineered bacteria).

◆ **Antibiotics** are natural antibacterial drugs, produced by moulds and bacteria. Penicillin was discovered (by Alexander Feming) in 1928, and isolated and concentrated for use as a drug (by Howard Florey and Ernst Chain) in 1940. Many more antibiotics have since been found.

◆ **Synthetic drugs** date from 1910 when the German chemist Paul Ehrlich developed an arsenic compound, arsphenamine (Salvarsan),

PENICILLIN
A drawing, with notes, of the original penicillin culture made by Sir Alexander Fleming.

to kill syphilis bacteria. This was the first chemical 'magic bullet'. The first antibacterial sulpha drugs, for treating infections, were produced in the 1930s.

◆ **Designer drugs** are 'designed' to interact with, and bind to, receptor sites on cells and natural substances in the body. Scientists model and 'test' the 3-D chemical structure of possible drugs on a computer.

◆ **Delivery systems**, apart from pills, include slow-release patches and capsules, injection guns, and small implantable pumps which deliver regular direct doses of a drug.

Surgery ❷

Lack of anaesthetics, loss of blood and the risk of infection made surgery agonising and dangerous until the mid 19th century.

◆ **Milestones** Before the development of proper anaesthetics, alcohol or opium was given before operations. The first use of **ether** was in 1842, of **nitrous oxide** in 1844, and of **chloroform** in 1847; Queen Victoria was given chloroform during childbirth in 1853. Today, injectable **anaesthetics** and halon gas are used. Joseph Lister pioneered **antiseptic** surgery, using carbolic acid (phenol) to clean wounds, in 1865. Later the emphasis was placed on

aseptic (sterile) surgery and thorough cleanliness to prevent any infection.

◆ **Modern surgery** Specialist techniques include minimally invasive ('**keyhole**') surgery, in which the surgeon uses an endoscope and a monitor as a guide. In **laser surgery**, lasers are used to cut, cauterise and 'weld' tissues. In **microsurgery**, the surgeon manipulates tiny instruments while looking through a binocular microscope. **Cryosurgery** uses extreme cold.

THROUGH THE KEYHOLE Surgeons performing laparoscope – or 'keyhole' – surgery.

WEIRD AND WONDERFUL ❸
Before the invention of anaesthetics in the mid 19th century, surgeons tried to work very fast – for the sake of their patients. In the early 1800s, one French surgeon could amputate a leg in just 15 seconds.

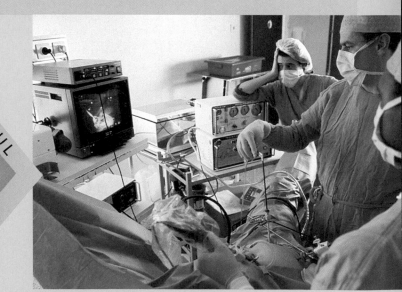

Medical machines

4

Advances in engineering, electronics and computers have led to the increasing use of machines in treating and monitoring the wellbeing of patients.

◆ **Iron lung**, invented in 1929, enabled polio patients to breathe even though their chest muscles were paralysed. Modern ventilators are much more compact machines.

◆ **Heart-lung machine**, first used in 1953, allows open-heart surgery by taking over the pumping and oxygenation of the patient's blood, so that the heart can be stopped.

◆ **Dialysis machines** take over from the kidneys when they fail. The first successful clinical machine to cleanse the blood directly was developed in the Netherlands in 1943.

◆ **Patient-monitoring** developed with the growth of intensive-care units from the late 1950s and particularly with microelectronics in the 1960s and 70s onwards.

Implants and transplants

5

Artificial body implants were crude until the mid-20th century. The first **hip-replacement** was carried out in 1938. An external **cardiac pacemaker** was first used in the 1930s, and the first successful implanted pacemaker in 1960. The first **artificial heart** was implanted in 1969, but living heart transplants have been much more successful.

A kidney was first transplanted in 1933, but success depended on better knowledge of tissue rejection and immunosuppression. The first successful **kidney transplant** was in 1954, the first **heart transplant** (by Christiaan Barnard) in 1967. Since then, tens of thousands of transplants have taken place.

MAINTAINING A REGULAR BEAT This coloured X-ray of a patient's chest shows a fitted heart pacemaker; the battery-run device is seen above the ribcage.

489

Barber-surgeons

In the Middle Ages, physicians (doctors) did not carry out surgery or blood-letting; they left this to barber-surgeons. Barbers advertised their medical role by tying bloody rags to poles outside their shops, from which came the traditional red-and-white barber's pole. This ancient separation of roles is why consultant surgeons are addressed as 'Mr', not 'Dr'.

Complementary medicine

6

Systems of treatment outside mainstream medicine are called **alternative** or, if practised in conjunction with conventional treatments, complementary medicine. They include ancient medical traditions such as Chinese **acupuncture**, Indian **Ayurvedic** medicine and Western **herbalism**.

Others are more recent, including **homoeopathy** (originated by Samuel Hahnemann in 1810), **osteopathy** (founded by Andrew Taylor Still in the 1870s) and **chiropractic** (begun by Daniel David Palmer in 1895). Many orthodox doctors today accept that complementary treatments may help certain conditions.

Preventive medicine

7

The greatest health improvements have come from preventing rather than treating disease. For example, Edward Jenner began smallpox **vaccination** in 1796. **Public health** measures are vital, too – as demonstrated in 1854 when John Snow traced a cholera epidemic to one water pump in London's Soho. The importance of **diet** was realised long before vitamins were discovered in the early 20th century. **Fluoride** was found to reduce tooth decay in the 1930s, and was first added to water supplies in the USA in 1945.

Contraception and conception

8

The biggest development in preventing conception came with the launch in 1960 of the oral contraceptive, or '**pill**'.

On the other hand, the most revolutionary development in the struggle to promote conception has involved fertilising egg cells outside the mother's body and then re-implanting them inside the womb. In vitro fertilisation or **IVF** (fertilisation in glass) was pioneered in England in the 1960s and 70s by Dr Patrick Steptoe and Dr Robert Edwards.

Science fact and fiction

Core facts ❶

◆ **Science fiction** is a genre of fiction, film, television and graphic art that makes use of imagined science and technology.
◆ The first **golden age** of science fiction came in the 1890s, with authors like Jules Verne and H.G. Wells.
◆ From the 1920s to the 1960s the genre was dominated by 'hard' science, from authors such as Robert Heinlein, Arthur C. Clarke and Isaac Asimov.
◆ **'Soft'** science fiction, concerned with social and cultural implications as much as with technology, began to appear in the 1960s from authors such as J.G. Ballard and Ursula Le Guin.

Scientific bestsellers

A number of **non-fiction** scientific books have become bestsellers, often unexpectedly. Some examples include:

Title	Author	Date	Comment
The Origin of Species	Charles Darwin	1859	The book that founded the science of evolution and effectively ended the creationist myth.
The Universe Around Us	James Jeans	1929	A major step in the popularisation of modern cosmology.
Silent Spring	Rachel Carson	1962	The book that almost single-handedly started the ecological movement.
The Double Helix	James Watson	1968	A bestselling account of the discovery of DNA, by one of the scientists involved.
The Right Stuff	Tom Wolfe	1980	A 'new journalism' account of the US space programme.
A Brief History of Time	Stephen Hawking	1988	An international bestseller outlining some of the most complex ideas in recent physics and cosmology.

A history of science fiction ❸

Title	Author	Date	Comment
Frankenstein	Mary Shelley	1818	The first science fiction novel, dealing with a human-created monster.
The Time Machine	H.G. Wells	1895	Classic account of time travel – and origin of an entire science fiction sub-genre.
Last and First Men	Olaf Stapledon	1930	Exploration of human and cosmological origins and futures, vast in scope.
I, Robot	Isaac Asimov	1950	Nine classic stories that established the robot as a science-fiction protagonist.
Do Androids Dream of Electric Sheep?	Philip K. Dick	1968	Moving, enigmatic study of the nature of humanity; basis of the film *Blade Runner*.
Neuromancer	William Gibson	1984	The most influential early 'Cyberpunk' novel.

ROBOT STORIES Isaac Asimov's *I, Robot* (1950) looks at how humans react to robots. Unlike many other robot stories of the time, Asimov's robots are helpful to society and are ultimately a tool of humanity.

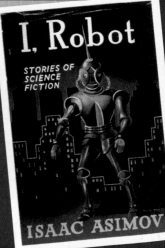

Dreams become reality

Science fiction has a mixed record in predicting **future science**. In some cases, such as travel to the Moon, authors like Jules Verne, in *From the Earth to the Moon* (1865), were right about the event, but wrong about the means – Verne envisaged a giant cannon rather than a rocket, though he predicted the crew's weightlessness. In others areas, such as contact with aliens, authors have been wildly wrong. Studies of the 'science' of popular science fiction like the TV show **Star Trek** have themselves been published.

CHANNEL TUNNEL Work finally began in 1986 (right), but early plans for a tunnel included this drawing by Herbert Clerget in 1875 (below).

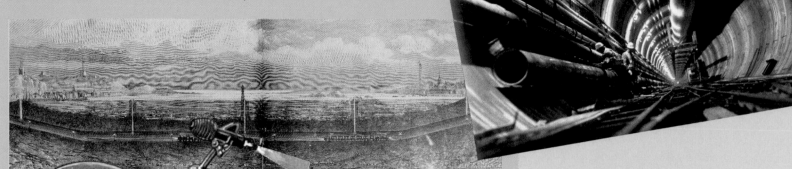

❹

MAGAZINE COVER A scene from Edmond Hamilton's 'The Reign of the Robots' featured on the cover of the December 1931 edition of *Wonder Stories*.

Scientists and inventors on the big screen

❺

With few exceptions, the cinema has been notably poor in its depiction of science and scientists. Clichéd '**mad scientist**' villains and **eccentric inventors** dominate screen images.

Title	Director	Date	Comment
Paris Qui Dort	René Clair	1923	An inventive comedy about a mad scientist who invents a ray that suspends animation throughout Paris.
Island of Lost Souls	Earl Kenton	1932	Charles Laughton as H.G. Wells's Dr Moreau, a vivisectionist; remade as *The Island of Dr Moreau* (1977, 1996).
Madame Curie	Mervyn LeRoy	1943	Greer Garson as the discoverer of radium; a surprisingly effective account.
Dr Strangelove	Stanley Kubrick	1963	Peter Sellers in a sharp satirical take on technological progress and Cold War paranoia.
Back to the Future	Robert Zemeckis	1985	Time-travel story featuring an inventor (Christopher Lloyd) whose time machine is actually a DeLorean sports car.
Insignificance	Nicolas Roeg	1985	Conversation piece set in a New York hotel in 1954; best scene has Marilyn Monroe (Theresa Russell) explaining relativity to Albert Einstein (Michael Emil).

★ **514**

Crazy ideas

Interest in wacky **inventions** as a form of entertainment has led to the creation of a hugely popular **Chindogu** ('weird tools') television show, website and publications by the Japanese comedian **Kenji Kawakami**. Typical examples include a hat with attached toilet roll for hayfever sufferers, and a portable zebra crossing.

WEIRD AND WONDERFUL

❻

A novel by **Jules Verne** was itself apparently stuck in a kind of **time warp**. His dystopian vision *Paris in the Twentieth Century* was rejected by his publisher in 1863, and first published in 1996, some 133 years later.

State-of-the-art science

Core facts ❶

◆ A great deal of current research into **computer technology** is focused on extreme miniaturisation, including the development of atom-sized components for **'quantum' computers**.

◆ In the field of **electronic communications** a truly high-speed **Internet** is in preparation, through which mobile computers and phones will connect people into ever more flexible and capable networks.

◆ **Medical research** is currently being transformed by the **Human Genome project**, which may prove the greatest breakthrough yet in the history of medicine.

New technology ❷

Most observers agree that the likeliest development in **computer** technology in the near future is to get rid of the computer: an entire computer may soon be contained in a mobile phone, or even a pair of glasses. **Biotechnology** and the Genome Project suggest that powerful 'designer drugs' may soon be available to treat most diseases.

FLYING MACHINE The Solo Trek XFV made its first piloted flight, just feet off the ground, in 2001. A more advanced prototype was due to be ready in 2003 for testing by US Special Forces.

⭐ **168**

Shrinking world

Miniaturisation is one of the clearest trends in technology, offering huge savings in materials and advances in mechanical flexibility. **Nanotechnology** – the development of microscopic and submicroscopic electronic and mechanical devices – is a major research area today. Scientists are already able to make components from individual atoms.

A virtual world ❸

Several current trends suggest that something like the **'Cyberspace'** of science-fiction writer William Gibson is becoming possible. As computing power continues to increase, so handling the huge amounts of data needed to create believable virtual environments is already feasible. Furthermore, such developments as electronic neurotransmitters – which can communicate directly with the human brain without intrusive wires – suggest that true **VR interfaces** may well become a reality.

VIRTUAL DRIVE Future cars are test-driven at the Daimler-Chrysler Virtual Reality Centre in Sindelfingen, Germany.

See hear ❹

The **mobile telephone** explosion is likely to continue, with technological advances vastly increasing the services available. Mobiles capable of receiving **streamed video** (visual material from the Internet) as well as sound are now entering the market. Future developments may include electronic banking, remote control of domestic equipment, fully voice-activated Internet access – and real-time translation for users speaking different languages.

POCKET PC This handheld PC is more powerful than the computers used to send men to the Moon in 1969.

Space technology ❺

Among probes planned or under consideration are the European Space Agency's **Mars Express**, scheduled for launch in June 2003; the ESA **New Horizons** probe to Pluto and Charon, for launch in 2006; the Japanese **Planet C** Venus orbiter, due for launch in February 2007; and NASA's **Europa Orbiter**, planned for launch in 2008.

Other plans include NASA's **Kepler** mission, a space-based project to detect potentially Earth-like planets in other parts of the galaxy.

Cleaning up ❻

Green technologies – such as sustainable energy generation, recycling, the development of cleaner and more efficient industrial plants, and the elimination of harmful chemicals – will grow in importance. One example nearing production is **biodegradable** packaging. Unlike the millions of tonnes of paper and plastic-based packaging that litters the landscape, biodegradable alternatives are made from starch rather than plastic and dissolve harmlessly after a few days' exposure to sun and rain.

SDI in the sky ❼

A political idea that will require major advances in several areas of technology is the further development of the US government's **Strategic Defense Initiative**, popularly known as Star Wars. The aim is the creation of a complete defence against intercontinental ballistic missiles by using **satellites**, **lasers** and **anti-missile missiles** to destroy them in the upper atmosphere.

Progress so far has been slow, expensive and politically divisive, and the value and true purpose of the system remain questionable. The challenges involved will have a great impact on future satellite, surveillance and weapons technology.

The numbers or star following the answers refer to information boxes on the right.

ANSWERS

29	**The Nobel prize for physics**
88	**Ivan Pavlov (1849-1936)**
133	**A: Dynamite** – patented in 1867
247	**The Ig Nobel prizes** – see (*) below
312	**Nobel prizes** – inaugurated in 1901
396	**A: William Bragg** – Physics 1915, 25 years old
430	**John Bardeen** – won Nobel prize in 1956 and 1972
538	**Germany** – lived 1858–1947
539	**Stockholm** – capital of Sweden
⭐ **649**	**20th century** ⭐
853	**The anniversary of Nobel's death**
876	**Marie Curie (1867–1934)**
949	**King of Sweden** – king of Norway presents Peace
987	**Marie Curie** – she won for a second time in 1911

(* for 'achievements that cannot or should not be reproduced.')

Nobel prize winners

649

Alfred Nobel

Alfred Nobel (1833-96) was a Swedish chemist and engineer. He invented **dynamite** (1867), **gelignite** (1875) and the smokeless gunpowder **ballistite** (1887). His manufacture of explosives, and development of **oilfields** in Azerbaijan, made him a large fortune, which he left in trust for the endowment of the prizes that bear his name. As well as **Chemistry** and **Physics**, Nobel prizes are offered for **Physiology or Medicine**, **Literature**, and **Peace**. Nobel refused to create a prize for Mathematics after his wife left him for a mathematician, but a prize for **Economics** was created in 1969.

Chemistry

1901 Jacobus van 't Hoff
1902 Hermann Fischer
1903 Svante Arrhenius
1904 William Ramsay
1905 **Adolf von Baeyer: development of organic dyes and hydroaromatic compounds**
1906 Henri Moissan
1907 Eduard Buchner
1908 **Ernest Rutherford: study of the chemistry of radioactive substances**
1909 Wilhelm Ostwald
1910 Otto Wallach
1911 **Marie Curie, née Sklodowska: discovery and study of radium and polonium**
1912 Victor Grignard, Paul Sabatier
1913 Alfred Werner
1914 Theodore Richards
1915 Richard Willstätter
1916 No prize awarded
1917 No prize awarded
1918 Fritz Haber
1919 No prize awarded
1920 Walther Nernst
1921 Frederick Soddy
1922 **Francis Aston: invention of mass spectrometry of radioactive isotopes**
1923 Fritz Pregl
1924 No prize awarded
1925 Richard Zsigmondy
1926 Theodor Svedberg
1927 Heinrich Wieland
1928 Adolf Windaus
1929 Arthur Harden, Hans von Euler-Chelpin
1930 Hans Fischer
1931 Carl Bosch, Friedrich Bergius
1932 Irving Langmuir
1933 No prize awarded
1934 **Harold Urey: discovery of deuterium (heavy hydrogen)**
1935 Frédéric Joliot, Irène Joliot-Curie
1936 Peter Debye
1937 **Norman Haworth, Paul Karrer: work on vitamins**
1938 Richard Kuhn
1939 Adolf Butenandt, Leopold Ruzicka
1940 No prize awarded
1941 No prize awarded
1942 No prize awarded
1943 George von Hevesy
1944 **Otto Hahn: discovery of nuclear fission**
1945 Artturi Virtanen
1946 James Sumner, John Northrop, Wendell Stanley
1947 Robert Robinson
1948 Arne Tiselius
1949 William Giauque
1950 Otto Diels, Kurt Alder

1951 Edwin McMillan, Glenn Seaborg
1952 Archer Martin, Richard Synge
1953 Hermann Staudinger
1954 **Linus Pauling: study of the nature of chemical bonds in complex substances**
1955 Vincent du Vigneaud
1956 Cyril Hinshelwood, Nikolay Semenov
1957 Alexander Todd
1958 **Frederick Sanger: study of the structure of proteins, especially insulin**
1959 Jaroslav Heyrovsky
1960 **Willard Libby: invention of radiocarbon dating**
1961 Melvin Calvin
1962 Max Perutz, John Kendrew
1963 Karl Ziegler, Giulio Natta
1964 Dorothy Crowfoot Hodgkin
1965 Robert Woodward
1966 Robert Mulliken
1967 Manfred Eigen, Ronald Norrish, George Porter
1968 Lars Onsager
1969 Derek Barton, Odd Hassel
1970 Luis Federico Leloir
1971 Gerhard Herzberg
1972 **Christian Anfinsen, Stanford Moore, William Stein: work on the structure of amino acids**
1973 Ernst Fischer, Geoffrey Wilkinson
1974 Paul Flory
1975 John Cornforth, Vladimir Prelog
1976 William Lipscomb
1977 Ilya Prigogine
1978 Peter Mitchell
1979 Herbert Brown, Georg Wittig
1980 **Paul Berg, Walter Gilbert, Frederick Sanger: work on nucleic acids**
1981 Kenichi Fukui, Roald Hoffmann
1982 Aaron Klug
1983 Henry Taube
1984 Bruce Merrifield
1985 Herbert Hauptman, Jerome Karle
1986 Dudley Herschbach, Yuan Lee, John Polanyi
1987 Donald Cram, Jean-Marie Lehn, Charles Pedersen
1988 Johann Deisenhofer, Robert Huber, Hartmut Michel
1989 **Sidney Altman, Thomas Cech: work on RNA (ribonucleic acid)**
1990 Elias James Corey
1991 Richard Ernst
1992 Rudolph Marcus
1993 **Kary Mullis, Michael Smith: development of the polymerase chain**

reaction, for duplicating strings of DNA
1994 George Olah
1995 **Paul Crutzen, Mario Molina, Sherwood Rowland: discovery of the destruction of the Ozone Layer by nitrogen oxide**
1996 Robert Curl, Harold Kroto, Richard Smalley
1997 Paul Boyer, John Walker, Jens Skou
1998 Walter Kohn, John Pople
1999 Ahmed Zewail
2000 **Alan Heeger, Alan MacDiarmid, Hideki Shirakawa: discovery of conductive polymers**
2001 William Knowles, Ryoji Noyori, Barry Sharpless
2002 John Fenn, Koichi Tanaka, Kurt Wuethrich

FIRST PRIZE The diploma awarded to Wilhelm Röntgen, who won the first Nobel prize for physics in 1901.

Physics

1901 Wilhelm Röntgen
1902 Hendrik Lorentz, Pieter Zeeman
1903 **Henri Becquerel, Pierre Curie, Marie Curie: research into radioactivity**
1904 Lord Rayleigh
1905 Philipp Lenard
1906 Joseph Thomson
1907 **Albert Michelson: for precision optical equipment and spectroscopic research**
1908 Gabriel Lippmann
1909 **Guglielmo Marconi, Karl Braun: development of wireless telegraphy**
1910 Johannes van der Waals
1911 Wilhelm Wien
1912 Nils Dalén
1913 Heike Kamerlingh-Onnes
1914 Max von Laue
1915 William Henry Bragg, William Lawrence Bragg
1916 No prize awarded
1917 Charles Barkla
1918 **Max Planck: discovery of energy quanta**
1919 Johannes Stark
1920 Charles Guillaume
1921 **Albert Einstein: theoretical physics and discovery of the photoelectric effect**
1922 **Niels Bohr: work on the structure of atoms**
1923 Robert Millikan
1924 Karl Siegbahn
1925 James Franck, Gustav Hertz
1926 Jean-Baptiste Perrin
1927 Arthur Compton, Charles Wilson
1928 Owen Richardson
1929 **Louis-Victor de Broglie: discovery of the wave nature of electrons**
1930 Chandrasekhara Raman
1931 No prize awarded
1932 **Werner Heisenberg: development of quantum mechanics**
1933 **Erwin Schrödinger, Paul Dirac: development of atomic theory**

NOBEL CENTENNIAL Professor Ryoji Noyori receives the Nobel prize for chemistry from King Carl XVI Gustav of Sweden in 2001.

1934 No prize awarded
1935 **James Chadwick: discovery of the neutron**
1936 Victor Hess, Carl Anderson
1937 Clinton Davisson, George Thomson
1938 Enrico Fermi
1939 Ernest Lawrence
1940 No prize awarded
1941 No prize awarded
1942 No prize awarded
1943 Otto Stern
1944 Isidor Rabi
1945 Wolfgang Pauli
1946 Percy Bridgman
1947 Edward Appleton
1948 Patrick Blackett
1949 **Hideki Yukawa: prediction of mesons**
1950 Cecil Powell
1951 John Cockcroft, Ernest Walton
1952 Felix Bloch, Edward Purcell
1953 Frits Zernike
1954 Max Born, Walther Bothe
1955 Willis Lamb, Polykarp Kusch
1956 **William Shockley, John Bardeen, Walter Brattain: development of the transistor**
1957 Chen Ning Yang, Tsung-Dao Lee
1958 Pavel Cherenkov, Ilya Frank, Igor Tamm
1959 Emilio Segrè, Owen Chamberlain
1960 Donald Glaser
1961 Robert Hofstadter, Rudolf Mössbauer
1962 Lev Landau
1963 Eugene Wigner, Maria Goeppert-Mayer, Hans Jensen
1964 Charles Townes, Nicolay Basov, Aleksandr Prokhorov
1965 **Sin-Itiro Tomonaga, Julian Schwinger, Richard Feynman: work in quantum electro-dynamics**
1966 Alfred Kastler
1967 Hans Bethe
1968 Luis Alvarez
1969 **Murray Gell-Mann: classification of elementary particles**
1970 Hannes Alfvén, Louis Néel
1971 **Dennis Gabor: development of holography**
1972 John Bardeen, Leon Cooper, John Schrieffer
1973 Leo Esaki, Ivar Giaever, Brian Josephson
1974 **Martin Ryle, Antony Hewish: research in astrophysics**
1975 Aage Bohr, Ben Mottelson, Leo Rainwater
1976 Burton Richter, Samuel Chao Chung Ting

1977 Philip Anderson, Nevill Mott, John van Vleck
1978 Pyotr Kapitsa, Arno Penzias, Robert Wilson
1979 **Sheldon Glashow, Abdus Salam, Steven Weinberg: theory of unified weak and electromagnetic interaction**
1980 James Cronin, Val Fitch
1981 Nicolaas Bloembergen, Arthur Schawlow, Kai Siegbahn
1982 Kenneth Wilson
1983 Subramanyan Chandrasekhar, William Fowler
1984 Carlo Rubbia, Simon van der Meer
1985 Klaus von Klitzing
1986 **Ernst Ruska, Gerd Binnig, Heinrich Rohrer: work in electron optics**
1987 Georg Bednorz, Alexander Müller
1988 Leon Lederman, Melvin Schwartz, Jack Steinberger
1989 Norman Ramsey, Hans Dehmelt, Wolfgang Paul
1990 Jerome Friedman, Henry Kendall, Richard Taylor
1991 Pierre-Gilles de Gennes
1992 Georges Charpak
1993 **Russell Hulse, Joseph Taylor: discovery of a new type of pulsar**
1994 Bertram Brockhouse, Clifford Shull
1995 Martin Perl, Frederick Reines
1996 David Lee, Douglas Osheroff, Robert Richardson
1997 Steven Chu, Claude Cohen-Tannoudji, William Phillips
1998 Robert Laughlin, Horst Störmer, Daniel Tsui
1999 **Gerardus 't Hooft, Martinus Veltman: work on the quantum structure of electroweak particles**
2000 Zhores Alferov, Herbert Kroemer, Jack Kilby
2001 Eric Cornell, Wolfgang Ketterle, Carl Wieman
2002 Raymond Davis Jr, Masatoshi Koshiba, Riccardo Giacconi

Famous laws and formulae

Core facts ❶

◆ Scientific laws are generalised rules – shown to be true by **experiment** – that describe defined natural phenomena.
◆ A law is only valuable if it relates to other laws on relevant phenomena.
◆ Many scientific laws set out mathematical relationships between phenomena as formulae.

◆ Apparent **exceptions** to scientific laws are generally useful for testing their validity; established laws are rarely disproven.
◆ A law that is incompletely tested by experiment is properly known as a **hypothesis**.
◆ **Thought experiments** are speculations that either test or illustrate scientific laws or ideas.

Laws of physics ❷

Most laws in physics apply to very narrow areas of the subject. Apart from **Newton**'s laws of motion and **Einstein**'s theories of relativity (see pages 54-5), the most widely applicable and familiar laws are probably the three **laws of thermodynamics**:

1 The energy of a thermodynamic system can be converted to other forms (such as work or heat), but its total remains constant – the principle of **conservation of energy**.

2 Within a thermodynamic system, the availability of energy for conversion declines inexorably over time – the principle of **increasing entropy**.

3 The entropy of a thermodynamic system is zero only at a temperature of absolute zero (0 K or zero kelvin).

The British writer and physicist C.P. Snow paraphrased the laws as:
1. 'You can't win' (because energy is conserved whatever you do);
2. 'You can't break even' (because available energy always declines); and
3. 'You can't even get out of the game' (because absolute zero is unattainable).

The laws were formulated in the 19th and early 20th centuries, partly as a result of intense scientific attention to mechanical efficiency in machinery. They represented a final proof that the ancient quest for a source of perpetual motion was mistaken.

Unusually, the laws of thermodynamics were influential well beyond the world of physics. The perception that the Universe was, in effect, a single thermodynamic system increasing in entropy affected 20th-century ethics, literature and even religion, as well as science.

Laws and rules of electricity ❸

The three basic units of electricity are related by a simple formula. In practical terms, 1 watt of power is delivered by a current of 1 amp flowing across a component with a potential of 1 volt:

Watt = amp x volt

Other well-known laws of electricity are:

Coulomb's Law
The French physicist Charles Coulomb's law of 1785 states that the electrostatic force between two charged bodies is proportional to the product of the charges divided by the square of the distance between the bodies.

Ohm's Law
Published by the German physicist Georg Ohm in 1827, Ohm's Law states that the current flowing through a given resistance is equal to the total applied voltage divided by the resistance.

Lenz's Law
In 1833, German physicist Heinrich Lenz observed that an induced current forms a magnetic field opposing the movement that induces it.

Kirchhoff's Laws
German physicist Gustav Kirchhoff noted in 1845-6 that the sum of voltages around a loop in an electrical circuit is always zero, and that at any junction in a circuit the sum of currents arriving at any instant always equals that of currents leaving.

Laws of chemistry

Many laws have been formulated in relation to chemistry. Among the most widely known are those that apply to the behaviour and properties of gases:

Boyle's Law
Published by Irish chemist Robert Boyle in 1662, Boyle's Law states that the volume of a given mass of gas at a constant temperature is inversely proportional to its pressure.

Charles's Law
Charles's Law was published by the French chemist and physicist Jacques Charles in 1787 (though it is also known as Gay-Lussac's Law, after Joseph Gay-Lussac, who demonstrated it in 1802). A derivation of Boyle's Law, it states that the volume of a given mass of gas at a constant pressure is directly proportional to its absolute thermodynamic temperature.

Dalton's Law of Partial Pressure
Published by the British chemist John Dalton in 1801, Dalton's Law states that the pressure of any gas in a mixture of gases is equal to the pressure that it would exert if it occupied the same volume alone at the same temperature.

Avogadro's Law
The Italian physicist and chemist Amedeo Avogadro, who first defined the term 'molecule' in distinction to 'atom', published a hypothesis in 1811 now recognised as a gas law. It states that equal volumes of different gases at the same pressure and temperature contain the same number of molecules.

Graham's Law
Formulated by the British chemist Thomas Graham in 1829, Graham's Law states that the diffusion rate of a gas is inversely proportional to the square root of its density.

Moore's Law
In 1965, Gordon Moore, co-founder of the Intel corporation, noted that the number of transistors we can fit on a silicon chip (and thus the power of computer processors) doubled every two years. 'Moore's Law' – that computers will continue to improve dramatically – has become a truism of the computer industry.

REVOLUTIONARY THEORIES
Albert Einstein's famous equation, $E=mc^2$, related matter and energy in his Special Theory of Relativity. Together with his General Theory, it altered the way that scientists viewed the Universe.

Tie-breaker

Q: Which law states that a substance always contains the same elements in the same proportions regardless of how it is made?
A: The law of constant composition. For example, carbon dioxide molecules in car exhaust and in exhaled breath are exactly the same in composition.

Famous thought experiments

Maxwell's Demon
Scottish physicist James Clerk **Maxwell** described his 'demon' in 1871, as an illustration of the **second law of thermodynamics**. The demon is an imaginary creature guarding a hatch in the partition of a sealed, divided chamber. As gas molecules pass the hatch he opens it to allow faster-moving molecules to move to one side. In time, though the total energy remains the same, higher energy becomes concentrated on the side with the faster molecules. The imbalance means that energy (heat) is available for conversion to work. Maxwell's point was that the **counter-entropic** demon was impossible in the real world, because his activity would itself draw energy from the system, cancelling any gain.

Schrödinger's Cat
Austrian physicist Erwin **Schrödinger** described his 'cat' in 1935 as an illustration of uncertainties in **quantum physics**. A cat is sealed in a chamber with a mechanism that has a 50 per cent chance of killing it within a given time. At the end of the prescribed time, the cat is definitely either 100 per cent dead or 100 per cent alive. But how does the observer know which is true, until the chamber is opened? In quantum physics the cat must be treated as 50 per cent alive – even though this defies common sense.
Probabilities are used in a similar way to describe quantum particles.

WEIRD AND WONDERFUL
Drake's Equation (1961) purports to define the number of intelligent species in our galaxy. In fact, all but one of its quantities (the rate of star formation) are pure speculation – suggesting that perhaps there are none at all.

QUESTION NUMBER

The numbers or star following the answers refer to information boxes on the right.

ANSWERS

3	**Inch** – 2.54 cm
269	**False** – tonne = 1000 kg, ton = 1016.05 kg
346	**Ten** ❻
350	**The hertz** ❷
584	**0°F** – lowest temperature Fahrenheit could create
590	**A litre** ❹
610	**Troy** – weighing system named for Troyes, France
629	**1024** – not 1000 as the prefix would suggest
664	**The watt** ❷
★ **671**	**The yard** ★
673	**1 million** ❷ ❸
675	**The French Revolution** ❶ ❺
676	**Angles** ❷ ❹
677	**112** – 14 lb = 1 stone, 8 stones = 1 cwt
678	**5280** – 3 ft = 1 yd, 1760 yd = 1 mile
680	**1 metre** ❺
771	**Milligram** ❷ ❸
803	**Pascal** – mathematician, SI unit, computer language
848	**Kelvin** ❷ ❻

SI units

Core facts ❶

◆ **Standard units of measurement** have made science truly international. The system of agreed units began in 18th-century France.
◆ Approved units are known as **SI units** from the French name for the system, *Système internationale d'unités*.

◆ The system is maintained by the Bureau Internationale de Poids et Mesures (BIPM).
◆ SI unit **symbols** are used internationally, though spelling of their names varies.
◆ The system evolves in response to **changing requirements** and measurement technologies.

SI units ❷

The SI system has seven mutually independent **base units**:

Unit name	Symbol	Quantity measured
metre	m	length
kilogram	kg	mass
second	s	time
ampere	A	electric current
kelvin	K	thermodynamic temperature
candela	cd	luminous intensity
mole	mol	amount of substance

The **metre** and **kilogram** were devised in 1799 and formally adopted by the BIPM in 1946, along with the **second** and **ampere**. The **kelvin** and **candela** were adopted in 1954, and the **mole** in 1971.

Twelve other SI units are derived directly from these base units, including units for area (**square metre**, m²), volume (**cubic metre**, m³), velocity (**metre per second**, m/s) and specific volume (**cubic metre per kilogram**, m³/kg).

A further 22 units are also derived from the base units, but have their own names and symbols. The relationship between these derived units and the SI base units can be mathematically complex: the **ohm**, for example, is defined as $m^2/kg/s^{-3}/A^{-2}$. Some of the more familiar examples of such derived units include:

Unit name	Symbol	Quantity measured
becquerel	Bq	radioactivity
degree Celsius	°C	Celsius temperature
hertz	Hz	frequency
joule	J	energy or work
newton	N	force
ohm	Ω	electric resistance
pascal	Pa	pressure or stress
radian	rad	plane angle
volt	V	electric potential difference
watt	W	power

A final series of units is derived from these SI units with their own names. Among these are units for surface tension (**newton per metre**, N/m) and angular velocity (**radian per second**, rad/s).

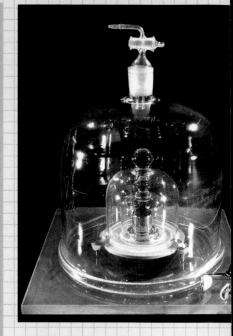

PRESERVED WEIGHT
The international standard kilogram prototype, kept at the BIPM since 1889.

SI prefixes ❸

Multiples of units are shown by prefixes:

exa	E	10^{18}
peta	P	10^{15}
tera	T	10^{12}
giga	G	10^{9}
mega	M	10^{6}
kilo	k	10^{3} (1000)
hecto	h	10^{2} (100)
deca	da	10
deci	d	10^{-1} ($1/10$)
centi	c	10^{-2} ($1/100$)
milli	m	10^{-3} ($1/1000$)
micro	μ	10^{-6}
nano	n	10^{-9}
pico	p	10^{-12}
femto	f	10^{-15}
atto	a	10^{-18}

Acceptable non-SI units ❹

Thirteen non-SI units are currently accepted for use with SI units, including the time measurements **minute** (min), **hour** (h) and **day** (d), the angle measurements **degree** (°), **minute** (') and **second** ("), the volume measurement **litre** (L) and the mass measurement **metric ton** or **tonne** (t). These are all definable in SI units – for example, 1 minute = 60 seconds (seconds being the SI base unit for time).

A range of other units is accepted for historical reasons for use with SI units, but should always be defined when used. Among these are the **nautical mile**, **knot**, **hectare** (ha), **ångström** (Å), **curie** (Ci) and **roentgen** (R). These are also definable in SI units – for example, 1 nautical mile = 1852 m.

WEIRD AND WONDERFUL ❺

The **metre**, when established in 1799, was intended to be one ten-millionth of a line running from the North Pole to the Equator. Modern precision measurements show this would have given a unit just 0.2 mm longer than a metre.

★ 671

Standards

Many devices have been used to standardise measurements. According to one 12th-century English chronicler, the **yard** was standardised as the length of King Henry I's arm. Later statutes laid down that an **inch** equalled the length of three barleycorns, and the **rod** (16½ ft) was defined as the length of the left feet of 16 men lined up coming out of church.

IMPERIAL SYSTEM
A display in a wall in Trafalgar Square, London, shows the traditional Imperial measurements: the inch, the foot, the yard and the chain (66 ft).

Tie-breaker

Q: Which temperature scale was originally devised with 0° as boiling point and 100° as freezing point?
A: Celsius. Devised by Swedish astronomer Anders Celsius (1701-44), the scale was reversed in 1948 (freezing point = 0°, boiling point = 100°).

Making conversions ❻

In some cases, non-SI units are widely used in everyday life alongside their SI equivalents.

For example, five different scales have been used to measure temperature – in historical order, Fahrenheit, Réaumur, Celsius, Kelvin and Rankine. Of these, the **Fahrenheit** scale remains in common use in Britain and the USA, although the common SI-derived unit for temperature is the degree **Celsius**.

The degree **Kelvin** (1 K), used by scientists, equals one degree Celsius (1°C), but the Kelvin scale has a different zero point: absolute zero (approximately –273.15°C, see page 60).

The **formulae for conversion** between Fahrenheit and Celsius are:
$$°C = (°F - 32) \times \tfrac{5}{9}$$
$$°F = (°C \times 1.8) + 32$$

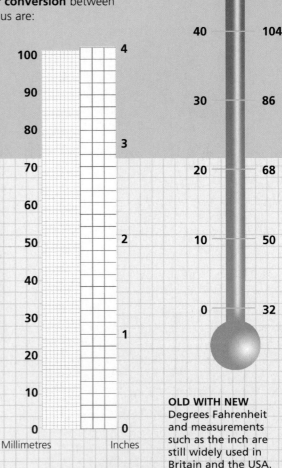

OLD WITH NEW
Degrees Fahrenheit and measurements such as the inch are still widely used in Britain and the USA.

The page number in brackets indicates where you will find the questions for each quiz.

Quiz 0 (page 8)

1 Indigo
2 Internet
3 Inch (2.54cm)
4 Infinity
5 Industrial revolution
6 Infrared
7 Inoculate
8 India
9 Insulin
10 Inorganic

Quiz 1 (page 8)

11 The first kidney transplant
12 The stethoscope
13 Hippocrates
14 AB
15 Penicillin
16 Laughing gas
17 Pox
18 The heart
19 Florence
20 Poppy

Quiz 2 (page 8)

21 Hubble Space Telescope
22 54
23 Harrods
24 Eight
25 The roof
26 Three Mile Island
27 A comet
28 To bleach it
29 The Nobel prize for physics
30 Harley-Davidson

Quiz 3 (page 8)

31 Diameter
32 Fraction
33 Decimal
34 Average
35 Percentage
36 Arithmetic
37 Pythagoras
38 Probability
39 Prime number
40 Perpendicular

Quiz 4 (page 9)

41 James Bond
42 The Michelin man
43 Motor town
44 A Volkswagen Beetle
45 Honda
46 The 1890s
47 Chevaux
48 The Edsel
49 Cadillac
50 In the exhaust system

Quiz 5 (page 9)

51 Albert Einstein's
52 DNA
53 *Frankenstein*
54 Stephen Hawking
55 Desmond Morris
56 Richard Dawkins
57 *The War of the Worlds*
58 Gaia
59 *Silent Spring*
60 Sir Thomas More

Quiz 6 (page 9)

61 A
62 A
63 C
64 D
65 A
66 B
67 A
68 A
69 A
70 C

Quiz 7 (page 10)

71 Algebra
72 Kodak
73 Neon
74 Stylus
75 Area
76 Bulb
77 Xerox
78 Refrigerator
79 Bomb
80 Nylon

Quiz 8 (page 10)

81 Intel
82 Blue
83 Knife or dagger
84 Windmills
85 Molecules are made of atoms
86 Glass
87 Stealth
88 Ivan Pavlov
89 Asbestos
90 Neutron

Quiz 9 (page 10)

91 Two
92 Two
93 The lower number
94 Myriad
95 Calculus
96 1000
97 Binary, or base 2
98 42
99 One-third
100 90

Quiz 10 (page 10)

101 She became the first woman in space
102 *Apollo 13*
103 A solar eclipse
104 Pluto
105 Russian
106 The Sun
107 Extraterrestrial intelligence
108 Mars
109 *Alien*
110 John Glenn

Quiz 11 (page 11)

111 Tower of London
112 A pyramid
113 An aqueduct
114 The 18th century
115 Newcastle-upon-Tyne
116 Cantilever
117 The Eiffel Tower
118 The Great Pyramid
119 Denmark and Sweden
120 Chicago

Quiz 12 (page 11)

121 Chlorophyll
122 Oxygen
123 Nitrogen
124 They are both antinuclear signs
125 A sundial
126 Coal
127 California
128 Hubble Space Telescope
129 The world's first solar calculator
130 Helium

Quiz 13 (page 11)

131 D
132 A
133 A
134 C
135 D
136 B
137 B
138 C
139 C
140 B

Quiz 14 (page 12)

141 False
142 False
143 True
144 True
145 True
146 False
147 True
148 True
149 True
150 False

Quiz 15 (page 12)

151 Windmill
152 Meltdown
153 Chain reaction
154 Geothermal
155 Plutonium
156 Dynamo
157 Turbines
158 Radioactive waste
159 Nuclear fission
160 Cooling towers

Quiz 16 (page 12)

161 Blood cells
162 Friction
163 The Earth's
164 By being born identical twins
165 Oxygen
166 The letter A
167 Africa
168 Nanotechnology
169 Photographs
170 Neutrons

Quiz 17 (page 12)

171 The circumference of a circle
172 Pythagoras's
173 Carbon dioxide
174 Boyle's Law
175 Its area
176 Isaac Newton
177 Resistance
178 75°, 105° and 105°
179 Their wavelength
180 Avogadro's Law

Quiz 18 (page 13)

181 Arecibo telescope
182 London Eye
183 Empire State Building
184 Sydney Harbour Bridge
185 Petronas Towers
186 Iron Bridge
187 Akashi-Kaikyo Bridge
188 Millennium Dome
189 Eiffel Tower
190 CN Tower

Quiz 19 (page 14)

191 Cuba
192 Mace
193 China
194 The colours of the rainbow
195 A machine gun
196 Germany
197 The Red Baron
198 Arrows
199 Sword
200 Scud missiles

Quiz 20 (page 14)

201 *The Terminator*
202 A heart
203 Kryten
204 *Artificial Intelligence:AI*
205 A dog
206 K9
207 Isaac Asimov
208 RoboCop
209 Marvin
210 An artificial heart

Quiz 21 (page 14)

211 Dolly
212 Harrow
213 Hybrid
214 Hydroponics
215 DNA
216 Organically
217 DDT
218 The combine harvester
219 A fish farm
220 First plant to be genetically decoded

Quiz 22 (page 14)

221 Wurlitzer
222 Thomas Edison
223 It was a wind-up radio
224 Stereo
225 The compact disc
226 Karaoke
227 Nipper
228 *High Fidelity*
229 Two
230 The 'A'

Quiz 23 (page 15)

231 *Titanic*
232 The Moon
233 A trimaran
234 Ultraviolet waves
235 Microwaves
236 Frequency
237 Red
238 X-rays
239 Gamma rays
240 Radio waves

Quiz 24 (page 15)

241 His wife's hand
242 Cyberspace
243 Runners not listed have odds of 50-1 or greater
244 *Snow White and the Seven Dwarves*
245 Isaac Newton
246 Galileo Galilei
247 The Ig Nobel Prizes
248 First full-length Internet film
249 The left and right-hand pages
250 Carbon dioxide

Quiz 25 (page 15)

251 B
252 A
253 B
254 C
255 A
256 A
257 A
258 B
259 D
260 D

Quiz 26 (page 16)

261 True
262 True
263 False
264 True
265 False
266 False
267 True
268 True
269 False
270 True

Quiz 27 (page 16)

271 Crossbow
272 Sabre
273 Musket
274 Submachine gun
275 Torpedo
276 Flamethrower
277 Sidewinder
278 Smart weapons
279 Howitzer
280 Trebuchet

Quiz 28 (page 16)

281 Finland
282 The 7 key
283 Have a nice day
284 Short Message Service
285 Australia
286 Facsimile
287 Will you go out with me?
288 Alexander Graham Bell
289 Wireless Application Protocol
290 The 19th century

Quiz 29 (page 16)

291 Red, blue, green
292 The BBC
293 Telstar
294 The 1960s
295 Betamax
296 Electrons
297 Al Jolson
298 Outside broadcast
299 PAL
300 First person on TV

Quiz 30 (page 17)

301 Hydrogen
302 The USA
303 In space
304 Morse code
305 *Gattaca*
306 Planets
307 Thomas Edison
308 35 gallons
309 Microwaves
310 Isotopes

Quiz 31 (page 17)

311 Printers
312 Nobel prizes
313 Bases
314 Energy
315 Ships
316 Sugars
317 Electric motors
318 Printing processes
319 Organelles
320 Virtual reality systems

Quiz 32 (page 17)

321 C
322 A
323 C
324 A
325 D
326 B
327 B
328 B
329 D
330 A

Quiz 33 (page 18)

331 California
332 Maize
333 Louis Pasteur
334 Whipped cream
335 Saccharine
336 Colonel Sanders'
337 C
338 The 19th century
339 Vanilla
340 First GM food sold

Quiz 34 (page 18)

341 Acid
342 How to make fire
343 Two-stroke engine
344 Arthur C. Clarke
345 Camber
346 Ten
347 Little Boy
348 The passenger lift
349 Police speed checks
350 The hertz

Quiz 35 (page 18)

351 A telephoto lens
352 Polaroid
353 A flash
354 The sprocket
355 David Bailey
356 Ultrasound
357 Digital cameras
358 A horse
359 The aperture
360 Camera obscura

Quiz 36 (page 18)

361 A bicycle
362 France
363 A monorail
364 Because hydrogen is highly flammable
365 An aircraft carrier
366 The hovercraft
367 Antifreeze
368 The Space Shuttle
369 Seaplanes
370 A sailing ship

Quiz 37 (page 19)

371 Static electricity
372 On the sea
373 The second
374 Six times
375 Reaction
376 On the Moon
377 It points to magnetic north
378 Electromagnetism
379 250 grams
380 The becquerel

Quiz 38 (page 19)

381 Modem
382 Magnesium
383 Radar
384 Evaporate
385 Alpha
386 Nitrogen
387 Diamond
388 Repeater
389 Isaac Newton
390 Momentum

Quiz 39 (page 19)

391 A
392 D
393 D
394 B
395 A
396 A
397 C
398 B
399 B
400 D

Quiz 40 (page 20)

401 Memory
402 Two metres square
403 Lead
404 Carbon
405 A cosmonaut comes from the former USSR
406 Apple
407 A ground floor
408 Metals
409 Men
410 Weightlessness

Quiz 41 (page 20)

411 Concorde
412 California
413 Henry Ford
414 Gliders
415 The Fat Controller
416 Indian Railways
417 Horses
418 Leslie Nielsen
419 Average speed
420 The *Enola Gay*

Quiz 42 (page 20)

421 Barrel
422 Barometer
423 Bar code
424 Barge
425 Barrage
426 Barbican
427 Barbiturates
428 Barnard (Christiaan)
429 Barium
430 Bardeen (John)

Quiz 43 (page 20)

431 Franklin
432 Cousteau
433 Galvani
434 Fermat
435 Maiman
436 Pilkington
437 Vesalius
438 Heisenberg
439 Landsteiner
440 Oersted

Quiz 44 (page 21)

441 Parallelogram
442 Nonagon
443 Ellipse
444 Scalene triangle
445 Trapezium
446 Heptagon
447 Kite
448 Rhombus
449 Isosceles triangle
450 Pentagon

Quiz 45 (page 22)

451 *Casino*
452 MTV
453 *Metropolis*
454 *Charlie and the Chocolate Factory*
455 Fonts or typefaces
456 The Moon
457 SOS in Morse code
458 *Tron*
459 Dirk Bogarde
460 Virtual reality

Quiz 46 (page 22)

461 Calcium
462 Very low temperatures
463 Because it is light
464 Iron
465 Mercury
466 Pewter
467 Copper
468 Zinc
469 Aluminium, nickel and cobalt
470 Aluminium

Quiz 47 (page 22)

471 The Beatles
472 The Jam
473 Prince
474 Radiohead
475 Pink Floyd
476 Battersea Power Station
477 The video for *Two Tribes*
478 The Prodigy
479 Freddie Mercury
480 Rod Stewart

Quiz 48 (page 22)

481 Leonardo da Vinci
482 Bill Gates
483 Archimedes
484 Eight bits
485 Carbon monoxide
486 Lead
487 Schrödinger's
488 World's largest ship
489 Blood and bandages
490 Computer Aided Manufacturing

Quiz 49 (page 23)

491 Bakelite
492 Tax on plastic bags
493 Celluloid
494 Oil
495 The nose
496 Lycra
497 Monomers
498 Rubber
499 Australia
500 Perspex

Quiz 50 (page 23)

501 6
502 Leonardo Fibonacci
503 Irrational numbers
504 100
505 The Simpsons
506 7 cm
507 Zero
508 The Golden Section
509 Ordinal numbers
510 A quarter

Quiz 51 (page 23)

511 D
512 A
513 D
514 B
515 C
516 D
517 D
518 B
519 D
520 B

Quiz 52 (page 24)

521 Pottery
522 Rust
523 The *Star Wars* films
524 A chromosone
525 The Bronze Age
526 Chemistry
527 Kevlar
528 Nitrogen
529 50:50
530 Organs

Quiz 53 (page 24)

531 Netherlands
532 Born in Germany and died in the USA
533 Russia
534 India
535 Japan
536 *Glue*
537 The USA
538 Germany
539 Stockholm
540 Austrian

Quiz 54 (page 24)

541 Renewable
542 Electricity
543 Nuclear
544 Sound barrier
545 Kilojoule
546 Conduction
547 France
548 Germany
549 Combustion
550 Potential

Quiz 55 (page 24)

551 Italy and Switzerland
552 Fool's gold
553 The Romans
554 London
555 Uranium
556 The Euphrates
557 To extract precious metals
558 Iron
559 Nuclear waste
560 The Cullinan diamond

Quiz 56 (page 25)

561 Hubble Space Telescope
562 Refracting telescope
563 X-ray machine
564 Scanning electron microscope
565 Light microscope
566 Magnetic resonance imager
567 Radio telescope
568 Transmission electron microscope
569 Thermograph
570 Atomic force microscope

Quiz 57 (page 26)

571 Hoover
572 Smelling salts
573 Mickey Mouse
574 Vinegar
575 Windows
576 James Dyson
577 Nintendo
578 America Online
579 Charles Goodyear
580 PVC

Quiz 58 (page 26)

581 Jim Morrison
582 The *Titanic*
583 Petrol
584 0°F
585 Fossil fuels
586 Ballast tanks
587 Surface tension
588 Heavy water
589 The solute
590 A litre

Quiz 59 (page 26)

591 Pole vaulting
592 Foil
593 The javelin
594 Bronze
595 Orienteering
596 Steel
597 Ice hockey
598 Kinetic energy
599 At a horse-race
600 Sailing

Quiz 60 (page 26)

601 A relay
602 A virus
603 Polaris
604 The nucleus
605 Proton
606 *The Crucible*
607 Gin
608 Perpendicular
609 The Palladium
610 Troy

Quiz 61 (page 27)

611 Agent Orange
612 Red
613 Blue
614 Red Adair
615 Ruby
616 Chess
617 *Primary Colours*
618 Sulphur
619 Green
620 Orange

Quiz 62 (page 27)

621 Oarsmen
622 Axe
623 China
624 Atari
625 A space rocket taking off
626 The Pulitzer Prize
627 An inflammation
628 Gordon Moore
629 1024
630 Quarks

Quiz 63 (page 27)

631 D
632 C
633 B
634 D
635 A
636 D
637 B
638 A
639 B
640 B

Quiz 64 (page 28)

641 An abacus
642 France
643 Air speed
644 Liquid
645 High frequency
646 In a hot air balloon
647 Personal computer
648 An atom of tin
649 20th century
650 $E=mc^2$

Quiz 65 (page 28)

651 Queen
652 20th century
653 'Marakesh Express'
654 Brass
655 A ferry
656 Stereo sound
657 A CB radio user
658 Microwave ovens
659 Marconi
660 Web radio stations

Quiz 66 (page 28)

661 China
662 Wood
663 The Beverley Hillbillies
664 The watt
665 OPEC
666 Nothing
667 Coal
668 The USA
669 Nickel and cadmium
670 A three-bar electric fire

Quiz 67 (page 28)

671 The yard
672 Your weight
673 1 million
674 Acceleration
675 The French Revolution
676 Angles
677 112
678 5280
679 Specific gravity
680 1 metre

Quiz 68 (page 29)

681 Virtual reality
682 Jetpack
683 Robotics
684 Genetic engineering
685 Hydrogen
686 Disposable
687 Biodegradable
688 Bluetooth
689 Strategic
690 International space station

Quiz 69 (page 29)

691 Weight
692 Numbers
693 Mensa
694 451
695 128
696 Byte
697 Base 60
698 500
699 Two
700 The Gutenberg Bible

Quiz 70 (page 29)

701 C
702 D
703 A
704 C
705 A
706 B
707 D
708 A
709 C
710 A

Quiz 71 (page 30)

711 The Gameboy
712 The pocket calculator
713 Louis Pasteur
714 The microscope
715 The Walkman
716 The Mini
717 A chip
718 Contact lenses
719 Mobile phones
720 Bacteria

Quiz 72 (page 30)

721 World Wide Web
722 Search engines
723 BBC
724 Hyperlink
725 Amazon.com
726 Sandra Bullock
727 Internet
728 The Vatican's
729 org
730 Its code was .tv

Quiz 73 (page 30)

731 The Trans-Siberian Railway
732 Roll-on, roll-off
733 'Buzz' Aldrin
734 Osaka
735 Water vapour
736 The Atlantic Ocean
737 On the undercarriage
738 On a hill or mountainside
739 The jet engine
740 To avoid both getting food poisoning

Quiz 74 (page 30)

741 The 'Tommy gun'
742 Indigo
743 Mercury
744 Switzerland
745 Oil fields (North Sea)
746 Half-Life
747 Artificial intelligence
748 Andrew Wiles
749 Neutrons
750 It doesn't

Quiz 75 (page 31)

751 1901
752 1938
753 1971
754 1876
755 c.1810
756 1930
757 1949
758 1888
759 2nd century BC
760 1st century AD

Quiz 76 (page 32)

761 The *Star Wars* films
762 It expands
763 Radar
764 The *Hindenburg* airship
765 Current
766 A human clone
767 Mobile phones
768 Douglas Adams
769 Pi
770 Publishing

Quiz 77 (page 32)

771 Milligram
772 Kayak
773 Tangent
774 Reactor
775 Asia
776 Metabolism
777 Neutron
778 Nineteen
779 Electrode
780 Lintel

Quiz 78 (page 32)

781 Greenhouse gases
782 Decomposes
783 Tissue
784 Nitrogen
785 Adaptation
786 Photosynthesis
787 Chromosome
788 Reproduction
789 Carbohydrates
790 Cytoplasm

Quiz 79 (page 32)

791 Hydrogen, oxygen
792 Carbon, hydrogen, oxygen
793 Energy
794 DNA
795 Catalyst
796 A covalent bond
797 Proteins
798 Methane
799 Sodium hydroxide and hydrogen
800 Oxygen

Quiz 80 (page 33)

801 Pacemaker
802 Quartz
803 Pascal
804 Quiver
805 Pager
806 Physics
807 Quorn
808 Phosphorus
809 Photon
810 Quanta

Quiz 81 (page 33)

811 Plants
812 Watches
813 Henry Ford
814 Nuclear fusion
815 Making pottery
816 Copper
817 The spinning jenny
818 The philosopher's stone
819 Cellophane
820 9000 years old

Quiz 82 (page 33)

821 D
822 A
823 D
824 B
825 B
826 B
827 C
828 C
829 A
830 D

Quiz 83 (page 34)

831 False
832 False
833 True
834 True
835 True
836 False
837 True
838 True
839 True
840 False

Quiz 84 (page 34)

841 Salt water
842 Flint
843 Metal
844 A microwave oven
845 Reykjavik
846 Alcohol
847 Metallic elements
848 Kelvin
849 Same in Celsius and Fahrenheit
850 Anneal

Quiz 85 (page 34)

851 1666
852 February 29th
853 The anniversary of Nobel's death
854 Nuclear power
855 January 6th
856 First global circumnavigation
857 The English Channel
858 The Islamic calendar
859 The 1910s
860 A lottery

Quiz 86 (page 34)

861 Underground railways
862 Frequency modulation
863 Ernest Rutherford
864 Plasma
865 First jet aeroplane
866 Sonar
867 The Czech Republic
868 Biomass
869 90 per cent
870 Marie Curie and Irène Joliot Curie

Quiz 87 (page 35)

871 Archimedes
872 Guglielmo Marconi
873 Albert Einstein
874 Craig Venter
875 Tim Berners-Lee
876 Marie Curie
877 Galileo Galilei
878 Isaac Newton
879 Edwin Hubble
880 Rachel Carson

Quiz 88 (page 36)

881 Mathematics
882 Botany
883 Microbiology
884 Polymer chemistry
885 Genetics
886 Biochemistry
887 Ecology
888 Climatology
889 Seismology
890 Thermodynamics

Quiz 89 (page 36)

891 Record deck
892 Epsilon
893 Microbiology
894 Wood
895 Carbohydrate
896 Harley-Davidson
897 Cotton
898 Barker
899 Television
900 E-type

Quiz 90 (page 36)

901 Eat with it
902 A food colouring
903 Non-stick pans
904 Andy Warhol
905 Because CFCs destroy ozone in the ozone layer
906 Iron
907 Monosodium glutamate
908 Botulism
909 Ice
910 Crabsticks

Quiz 91 (page 36)

911 9
912 3
913 The foxglove
914 Digital Versatile Disc
915 The compact disc
916 A picture file
917 Pixels
918 Toy Story
919 On a screen
920 Machine code

Quiz 92 (page 37)

921 Back to the Future
922 Thursday
923 Harvest moon
924 It enables them to date objects precisely
925 A water clock
926 Venus
927 The pendulum
928 The stars
929 1pm
930 10 000 years

Quiz 93 (page 37)

931 Internet
932 Telegram
933 Scotland
934 Satellites
935 Transceiver
936 Cellular
937 Fibre-optic
938 Third generation
939 Integrated
940 Wheatstone

Quiz 94 (page 37)

941 Convection
942 Nintendo
943 They all have nuclear weapons
944 Man of the Year
945 A sling
946 Sulphuric acid
947 Geordies
948 The computer user
949 The King of Sweden
950 Short 'tail lines'

Quiz 95 (page 37)

951 Light bulb
952 Ampère
953 The metal detector
954 Commutator
955 Transistor
956 Capacitors
957 50 ohms
958 The size of fuse required
959 Electrons
960 Solenoid

Quiz 96 (page 38)

961 Subtraction
962 Iron
963 Cotton
964 Silicon
965 Carbon
966 Radon
967 Irrigation
968 Newton
969 Population
970 Respiration

Quiz 97 (page 38)

971 HAL
972 The atomic bomb
973 The shield
974 Albert Einstein's
975 Coins
976 In the nucleus
977 France
978 Detecting submarines
979 The 1960s
980 The neutron

Quiz 98 (page 38)

981 The common cold
982 Fluorides
983 Chloroform
984 Aspirin
985 She was the first test-tube baby ever born
986 Polio
987 Marie Curie
988 Hermann Rorschach
989 Goitre
990 Electrocardiogram

Quiz 99 (page 38)

991 Mir space station
992 James T. Kirk
993 Voyager
994 Mars
995 Five
996 Weather satellites
997 Venus
998 It moved closer to the Sun than Neptune is
999 Io
1000 Pencils

Question sheet

Quiz Number	Quiz Title

Questions	Answers
1	1
2	2
3	3
4	4
5	5
6	6
7	7
8	8
9	9
10	10

Answer sheet

Name

Quiz Number	Quiz Title

Answers

1

2

3

4

5

6

7

8

9

10

Total score

The World of Science and Technology was published by The Reader's Digest Association Ltd, London. It was created and produced for Reader's Digest by Toucan Books Ltd, London.

Questions set by
David Bodycombe

Reference section written by
Simon Hall
Michael Wright

For Toucan Books:
Editors
Jane Chapman
Helen Douglas-Cooper
Daniel Gilpin
Andrew Kerr-Jarrett
Picture researcher
Christine Vincent
Consultant
Steve Allman
Proofreader
Ken Vickery
Indexer
Dorothy Frame
Design
Bradbury and Williams

For Reader's Digest:
Project editor
Christine Noble
Project art editor
Jane McKenna
Pre-press accounts manager
Penny Grose

Reader's Digest, General Books:
Editorial director
Cortina Butler
Art director
Nick Clark

Colour origination
Colour Systems Ltd, London

Printed and bound
in Europe by Arvato Iberia

First edition Copyright © 2003

The Reader's Digest Association Ltd,
11 Westferry Circus,
Canary Wharf,
London E14 4HE
www.readersdigest.co.uk

Reprinted with amendments 2003

We are committed to both the quality of our products and the service we provide to our customers. We value your comments, so please feel free to contact us on 08705 113366 or via our web site at
www.readersdigest.co.uk

If you have any comments or suggestions about the content of our books, you can email us at
gbeditorial@readersdigest.co.uk

ISBN 0 276 42716 5
BOOK CODE 625-003-02
CONCEPT CODE UK0095/G/S